The Einstein Paradox

*Sherlock Holmes and Professor Challenger
were created by the late Sir Arthur Conan Doyle,
and appear in stories and novels by him.*

The Einstein Paradox

And Other Science Mysteries
Solved by Sherlock Holmes

Colin Bruce

HELIX BOOKS

PERSEUS BOOKS
Reading, Massachusetts

Library of Congress Card Number: 98-87055

ISBN 0-7382-0023-9

Originally published under the title *The Strange Case of Mrs. Hudson's Cat*

Perseus Books is a member of the Perseus Books Group

Cover design by Suzanne Heiser
Text design by Irving Perkins Associates
Set in 11-point Garamond Light by Pagesetters, Inc.

23456789-DOH-020100999897

Find Helix Books on the World Wide Web at
http://www.aw.com/gb/

Contents

Preface

At the end of the last century, fundamental science appeared to be approaching a triumphant completion. The Universe operated according to straightforward, intuitively understandable laws which had been precisely described. The great Lord Kelvin even suggested that future investigators might have to confine themselves to making ever more accurate determinations of the fundamental constants of physics: there was no new territory left to explore.

Yet there were a few anomalies to be resolved. One paradox concerned the speed of light, which appeared bafflingly constant whatever the motion of the source and the observer. Others involved the microscopic world, which seemed oddly resistant to precise description. In the early decades of this century, these unresolved details were to blow apart the comfortably exact picture of the Universe which nineteenth-century scientists had so patiently assembled. We have not recovered it yet. The paradoxes still outstanding are more fascinating than any brainteasers devised by human puzzle setters; yet they have in common the tantalizing feel that they could be solved by a clever intuitive leap.

I have two reasons for retelling the story in this rather unorthodox form. The first is that I sympathize with Watson's plea in these pages: "No mathematics, Holmes: I have a horror of algebra." I wanted to set out the apparent paradoxes of special relativity and quantum theory in purely visual and logical terms, so that every reader has a fair chance to think about them for him- or herself and make up his or her own mind as to whether there is any alternative to the bizarre description of Nature given by modern physicists. The second is to make the information as digestible as possible. When I walk into a bookshop nowadays, I am quite intimidated by the volume of science books on display. I would be a better man if I read this, I think, picking up some excellently informative tome. But I'm not a better man: I'm a lazier one, and so I pass on to the lighter sections of the shop. We are all overwhelmed by information nowadays, and I have tried my best to make the stories no harder to read than more innocent fiction.

I am most grateful to Dame Jean Conan Doyle for her permission to use her father's famous characters. Sir Arthur had a genius for the believable depiction of ultimately clever men: in the world's imagination, Sherlock Holmes has reigned triumphant in his field for over a century. We know that he regarded himself as a scientist. Many of his famous aphorisms—knowledge follows first from observation and then from deduction; don't theorize ahead of the facts; accept the improbable once the impossible has been eliminated; an exception disproves the rule and must not be ignored—describe just those rules that good scientific investigation should follow, in plain language that should be the envy of some modern philosophers of science. I have also borrowed the famously irascible and outspoken Professor Challenger. In a world where too many scientists are learning the ways of politicians—don't criticize your seniors' views; be polite and evasive when asked to comment on nonsense—we sorely need his like.

My thanks go also to my sister Belinda and my editor Jeff Robbins, who each read the manuscript as it progressed and made many valuable suggestions.

Oxford   Colin Bruce
April 1997

The Einstein Paradox

1

The Case of the Scientific Aristocrat

"I HOPE FOR YOUR SAKE, WATSON, that popular science does not rot the brain."

I looked up from the illustrated scientific magazine I had been reading. Sherlock Holmes lounged opposite me in the more comfortable of our armchairs, idle save for his pipe.

"I am trying to broaden my mind a little," I said with some asperity. "No doubt you consider me incapable of grasping the subtleties—"

"Not at all, Watson, perish the thought! I was merely about to express the hope that your article is not written in the style of a lecture, as is sometimes the case, but provides enough information to allow an intelligent reader such as yourself to apply his own mind to form his own opinions."

Holmes's eyes scanned the cover. "Which article are you in fact reading? The one on the nature of the stars? That on the origins of the Earth?"

I felt myself blushing. "Well, in fact, Holmes, the magazine is also currently serializing one of Mr. Herbert George Wells's works, *The Time Machine*, and I was just glancing—"

My friend snorted.

"Really, Holmes, I am trying my best!" I cried. "But if you are attempting to educate me, you must make some allowances. For one thing, the subtleties of mathematics are quite beyond me."

Holmes smiled, then raised his right hand solemnly with the palm toward me. "Agreed, Watson. You have my word upon it: whatever else I may seek to tax your brain with, definitely no mathematics. In fact, the mathematical details are usually secondary to the logic of scientific inquiry. It is understanding the principles, not the calculation, which is the important thing."

"My other problem, Holmes, is that the subject matter can really be a little dry. It is human affairs and conflicts that interest me, however petty they may be in the grand cosmic scheme."

"I quite agree with you there, Watson, as my own choice of work should tell you. But it is always possible—"

However, at that moment we were dramatically interrupted. We had been too involved in our conversation to take much note of noises below; now the door was flung open, and a smartly dressed but wild-eyed young man burst in upon us.

"Mr. Sherlock Holmes? You must help me, sir, I implore you. It is my father. We must go to him at once."

I sprang up. "He is ill?" I said.

"He is dead! And, sirs, the only person who could possibly have done it—the person the police will be certain has done it—is myself!"

Sherlock Holmes raised his eyebrows. "And who are you, pray?"

"I am Viscount Forleigh. My father is Lord Forleigh. You may well have heard of him. He has long been famous as a classical scholar, and more recently as a philanthropist. His great project is the Universal Planetarium, now nearing completion not two miles from here."

We both nodded. No resident of London could have missed seeing the imposing structure which had been under construction on the South Bank of the Thames. A huge dome of green glass, it was a match in size for St Paul's Cathedral opposite, although unkind critics had likened it rather to an overgrown railway station.

The building was to house an exhibition of scientific devices and models, for the edification of the public, but the precise nature of the exhibits had remained a well-kept secret.

"As you no doubt know," the Viscount continued, "my father wished the contents of the building to remain a surprise until its formal opening, scheduled for the day after tomorrow. To this end, the workmen have been sworn to secrecy, and there are only two keys to the main door of the building. One was in my father's possession, one in mine. The other entrances have been kept blocked up, so one of us must always be present at the start and finish of the day's work, to permit the artisans access and then lock up behind them.

"Today is of course a Saturday, but my father was at the Planetarium making a final inspection. He had asked me to join him there at noon. I arrived a few minutes early, and found the main door locked. I opened it to enter, then relocked it behind me. I went forward onto the floor and called out. There was no reply. I assumed my father had been delayed, and started to inspect the layout for myself. All appeared in order. Then I turned a corner, and lying face down upon the floor before me, the back of his head caved in by some powerful blow . . ."

He stopped, his shoulders shaking.

"Your father?" asked Holmes quietly.

"Yes. He had evidently been coshed, American gangster style, by someone who had been lying in wait for him. The shock was considerable, although I shall not conceal from you, Mr. Holmes, my father and I were no longer on entirely friendly terms. I had made no secret of the fact I considered he was squandering too much of the family fortune, my inheritance-to-be, on his various projects.

"And so it dawned upon me that things looked rather black for me. The lock to the building is of Swiss design, quite unpickable. Of course I checked the premises carefully, but there was no sign of forced entry, nor any intruder. The only person who could have entered the building besides my father was myself. My father's key, incidentally, was still on the chain round his neck, where he always kept it.

"I thought that if I went to the police, they would inevitably arrest me. And so I locked the dome again and came to you."

My companion rose, rubbing his hands. "A capital puzzle," he exclaimed. "I have dealt with locked-room murders before, but this is certainly my first locked-museum mystery. We will have to involve the authorities before long, but we can assuredly go there first and make our own inspection. As you say, the police can be unimaginative, and I would not wish you to suffer the unpleasantness of detention."

DESPITE THE URGENCY, our client insisted on first calling past University College to let the project's chief scientific adviser, one Professor Summerlee, know what had occurred. But Summerlee was not to be found, and further time was consumed in writing a note for him. The fastest way from there to the Embankment was to take an Underground train from Euston. This we did, but there was some snarl-up with the signals, and we sat alongside another train in the darkened tunnel for several minutes, while our client grew ever more agitated.

"Ah. We are moving!" I exclaimed in relief, as the windows of the adjacent train started to slide past ours. No one contradicted me, but a few moments later the end passed us, and it suddenly became evident that we were in reality still stationary: it was the other train which had been in motion. I started to apologize for my error.

"It is an easy mistake to make," said the Viscount, "in the Universe at large as well as on a train. Indeed, since all the stars and planets traverse the skies at differing velocities, there is in a real sense no absolute state of rest or motion. You see Mars fly along at tens of kilometers per second, but a Martian observer would prefer to believe that he is at rest and the Earth is rushing headlong. We could even say you were right: the other train could be considered still stationary, and our own and the Earth in motion."

I thought the shock must have unhinged him slightly. The point might have some childish philosophical validity, but several million Londoners would agree unambiguously which train was at rest for practical purposes. But he continued:

"My father felt strongly about the equal validity of different viewpoints and world systems. He always considered that the ruthless reductionism of Western science, the idea that there was a clearcut right and wrong answer to every scientific question, was most unpalatable.

"He also felt we should have more respect for the views of the ancients. For example, Greek philosophy was in so many ways superior to our own: we should not despise them for lacking the accurate measuring instruments that would have given them our modern perspective on physics. They believed not in experiment but in logic: by discussion alone they found which hypotheses led to paradoxes and contradictions, and thus arrived at a rational view."

"I do not think the shortcomings of Greek philosophy can be blamed on inadequate instruments alone," said Holmes. He evidently felt that any conversation which distracted him from his plight would be good for our client.

"For example, take their notion that an object twice as heavy should fall twice as fast. Of course, we now know that, in the absence of air friction, all objects fall at the same rate, irrespective of size or density. Suppose some Greek philosopher had wanted to test this belief, without stooping to actual experiment. He might have imagined a system consisting of two bricks cemented together and dropped, falling at a given rate.

"Now, suppose that you were to file through the cement and drop the bricks side by side as before. Each brick would have just half the weight of the combined system. Would you expect each to fall half as fast as before?

"You could take the reductio ad absurdum further. Would it make a difference whether the two bricks were joined by a hair-fine thread? A ridiculous paradox! No, however clever the Greeks were at philosophy and politics, I fear we must accept their limitations as scientific thinkers."

"Well, my father was a very clever and renowned classicist, and I think I will respect his opinion over yours, Mr. Holmes," replied the Viscount, much put out.

My companion made no reply. He sat quiet for the remainder of the journey, and seemed to come fully alert again only when we stood before the great doors of the Planetarium itself. The Viscount ushered us inside, and turned a massive switch set in the wall.

I let out an involuntary gasp. Electric lights sprang on all over the ceiling in an irregular pattern. In a moment I saw that they formed the outlines of the familiar constellations. They were set not on black sky, however, but on a brilliantly painted mural which covered the dome. I recognized Diana the Huntress, the Lobster, the Crab: the ancient Greek constellations. The display was beautiful, yet there was something disconcertingly pagan about it.

The floor of the dome was covered with various machinery, all stationary, and walled constructions. The Viscount led us in toward the center. Holmes and I both tensed as a large object swung suddenly across our field of view.

"I beg your pardon, I did not mean to startle you," said the Viscount. "That is the great pendulum which symbolizes Time."

As we neared the center, we saw the pendulum clearly. It was suspended from the middle of the dome, fifty meters above our heads, and swung ponderously to and fro, passing waist high above the center of the floor. Near the center some brightly painted trestles had been scattered about on their sides: no doubt they would be arranged to form an aisle, to prevent bystanders from carelessly strolling into the pendulum's path when the Planetarium was opened to the public.

Some ten meters from the center of the dome was a most sinister sight. A man lay face down, feet toward the center, head away from it. The back of his head was matted with blood. I knelt beside him. It took only a few seconds to assure myself he had been dead for at least six hours.

We looked about us. There was quite an array of heavy machinery in sight, but none close enough to be a plausible source of accident. The obvious culprit, the pendulum, was swinging in completely the wrong direction: north–south, passing at least ten meters away from the body, which lay roughly east of the center. Even if the impact had flung the body some distance, the pendulum was swinging in the wrong direction to have been the cause.

Sherlock Holmes walked around unhurriedly, examining the various artifacts.

"This is a beautiful piece of work," he observed, stopping by a globe of the Earth some two meters in diameter. "Carved in relief— why, I can feel the Himalayas rising a good millimeter above the surface. And beautifully balanced." He turned the globe slightly on its gimbals. "However, its only motion is to rotate, so it can hardly have caused the accident. But what is this?"

Close to the globe was a circular table, mostly painted blue. Continent shapes were carved in relief on its surface. At the center a white cap topped a protruding pipe.

"That is the flat Earth, Mr. Holmes, as envisaged by the early Europeans," said the Viscount. "In operation, water is pumped from the hole at the center, streaming from beneath the ice at the North Pole—"

"And flows over the edge of the world, in a continuous all-round waterfall," finished Holmes. "It must be beautiful in operation."

"An absurd notion to contemplate, however. The flow of water would be somewhat noticeable," I could not resist commenting.

The Viscount spoke coldly. "Many peoples have believed in a flat Earth, Doctor, and who are you to deny the validity of their cultures? My father believed to his dying breath that the world view of, say, an American Indian or an Australian Aborigine had as much right to be respected as yours or mine."

As we walked on, Holmes spoke low in my ear. "If our client was voyaging at sea, short of rations, and found that his navigator believed the Earth was flat, and so was calculating the course wrongly, do you think he would be quite so generous? I think that might concentrate his mind wonderfully as to the validity of different world views! Ah, the conceit of the artistically minded aristocrat. But what have we here?"

He stood before an arrangement with a central capstan or gearbox from which horizontal arms of various lengths protruded in all directions. Each arm ended supporting a sphere of tinted glass, although the spheres differed greatly in size and coloring. A powerful electric bulb was mounted atop the capstan.

"Why, Watson, it is an orrery. See, the bulb at the center is the Sun. The shortest arm supports the red globe of Mercury. The blue-green one represents Earth. The huge multicolored ball—what glassblower did that, it is a feat!—is Jupiter. Saturn's rings are a bit of a give-away. That farthest one must be Neptune."

He inspected the apparatus more closely. "The central gear-box seems slightly more complex than in the other orreries I have seen," he commented.

"Yes, Mr. Holmes, this one takes account of the fact that the planetary orbits are ellipses rather than circles, and that each planet's speed varies inversely with its distance from the Sun," said Forleigh. "By a simple arrangement of off-center cogs, an extraordinary accuracy with respect to actual planetary movements is achieved."

To my astonishment, Holmes ducked into the arrangement, and stood up with his head inside the glass globe of the Earth: evidently, a hole had been cut in the base to make this possible.

"Quite right, Mr. Holmes: you are now seeing the planets just as they look from our Earth at this moment in time. You can similarly obtain a Martian or Jovian perspective if you wish. The orrery can be run forward or backward at high speed, to show the heavens as they have been or will be many thousands of years from now, so in a sense it is a time as well as a space travel machine."

"I hope you will be inviting Mr. Wells to your grand opening!" I said.

Holmes halted before the next exhibit in some perplexity. Superficially similar to the orrery, it featured at the center a large stool painted with the Earth's continents. From the stool ran arms which supported planetary globes. Each arm was multijointed, with a small gear-wheel arrangement at each joint. The device showed evidence of much tinkering, and was clearly not yet functional.

The Viscount appeared embarrassed. "That is an astrolabe, Mr. Holmes, but of more modern construction than those you usually see."

Holmes seated himself on the stool at the center. "Ah, of course.

The Epicycle Machine

And the view is indeed identical to that from the Earth globe on the orrery. An ingenious demonstration."

He turned to me. "You recall, Watson, that before the great astronomer Copernicus, it was believed that the Earth sat stationary at the center of the Universe. The celestial globe, to which were attached the fixed stars, turned about it once a day, and the planets, Moon, and Sun also orbited the Earth. The orbits were believed to be exactly circular, reflecting Divine perfection.

"Alas, even primitive measurements easily demonstrated that the apparent orbits of the planets about the Earth were nothing like circular. But the great theorist Ptolemy was able to patch the concept up, with the introduction of epicycles.

"He posited that each planet was turning in a perfect circle about an invisible fulcrum, which in turn was moving in a perfect circle about the Earth. Thus the combined motion could make the planet speed up or slow down at certain times, yet was still based on the 'perfect harmony' of uniform circular motion.

"More accurate measurements showed that a single epicycle per planet was not enough. You had to posit a circle about a point which moved in a circle about a point which moved in a circle . . . and so forth.

"However, the epicycle theory could never be formally disproved. By adding more epicycles, it always turned out to be possible to track the planetary motions to whatever accuracy was available from observation. But the epicycle concept became so complex and unwieldy that there was great need for a simpler system. Eventually the monk Copernicus posited that if you assumed all the planets, including the Earth, orbited the Sun, and that the Earth itself not only moved but also turned on its axis, a far simpler picture was possible."

"That sounds like a lot of new ideas to accept all at once," I said.

"It was indeed. Copernicus did not even try to convince people that the Earth *really* orbited the Sun. He simply suggested that if this was assumed purely as a mathematical convenience, like the arithmetical tricks often used to simplify difficult calculations, then celestial predictions could be made both more easily, and with more accurate results."

"A wise precaution."

"Quite so. When Galileo put the case for a Sun-centered system more bluntly, the Pope forced him to recant, on pain of torture by the Inquisitors. Any challenge to the knowledge and wisdom of the authorities was resented then as now. Even today the Catholic Church has yet to apologize or concede that Galileo and Copernicus were right. I do not wish to sound cynical, but perhaps in another century or so we may hope for a more liberal Pope who is equal to the task."

Viscount Forleigh coughed to attract our attention.

"The device you see before you is in fact the one of which my father was most proud. He felt that the old pre-Copernican view deserved its place in men's regard, and set out to devise a new system of epicycles.

"I am afraid our scientific adviser Professor Summerlee rather sneered at the machine. He could not deny that the device was showing fairly accurate results, but he kept pointing out small discrepancies which forced the addition of further epicycles, until the drive gears and wheels became impossibly numerous and finicky."

Holmes stood still, apparently in deep thought, as the Viscount paced nervously about. I pointed to the astrolabe in some excitement:

"Look at the arm holding Neptune," I said. "At the moment the segments are more or less folded together, but at some time they must all point out in the same direction."

I had a vision of the astrolabe being whirled toward some future date, or perhaps malfunctioning: the arm flying out, the globe impacting Lord Forleigh as he stood unawares.

"Hardly, Watson: the reach could never be sufficient. And I scarcely think the impact would leave the globe unmarked. No, I fear that personable as the son is, we must put technical red herrings aside and face the mundane evidence," said Holmes dryly. He advanced on the young Viscount.

"It is now several hours since the—accident—sir. I will continue to investigate if you wish, but I fear we must now proceed to the police and place the matter in their hands."

The Viscount's face turned white, but before he could speak, there came a furious hammering on the outside doors. He turned toward the sound.

"No, stay here, Viscount! Watson, take his keys. I believe our official colleagues are here," said Holmes quickly.

But on opening the door, I found myself facing a slim elderly man.

"I am Professor Summerlee, Director of this project. Let me in, I demand." With a swiftness that belied his age, he sidestepped me and went into the dome. He took in the scene in an instant.

"A Foucault pendulum!" he cried. "Ah, if only His Lordship saw fit to take professional advice, rather than relying on his flawed judgment."

He gestured imperiously at Holmes, who was holding the Viscount in a light grip, which I knew could be tightened instantly into a judo hold should the man try to escape. "Release him, sir: there has been no crime here."

He looked around at us. "If you release a pendulum," he said in the manner of one talking to idiot children, "it will continue to swing in the same direction in which it is initiated, will it not?"

"Why yes," I said, "and that is precisely why we eliminated it from our inquiries."

Summerlee snorted. "And the Earth—does that also maintain a constant orientation?"

Holmes cursed and smote himself on the forehead, but I doubtless looked as baffled as I felt.

"Take a simple case," said Summerlee, stepping over to the giant globe of the Earth. "I set a pendulum swinging to and fro at the North Pole. It swings in the direction from Pegasus to Virgo"—he had glanced briefly at the dome above—"and continues to do so. Now time passes, the Earth rotates—"

"The *apparent* direction of the pendulum changes," I exclaimed.

"Exactly so. In six hours, it is swinging at right angles to its original direction, as perceived by an insect standing on the surface." He peered in my direction, rather than at the globe. "The geometry is a little more complicated if the pendulum is located not at the Pole but at some intermediate latitude; however, the principle is the same: the relative direction of the pendulum swing changes as the globe turns.

"I knew that Lord Forleigh had some new idea for the dome. He had a childish habit of concealing details from me, possibly as a result of a little well-justified criticism that I gave from time to time. He must have had the pendulum rigged up last night, set it in motion, then returned this morning to inspect the other arrangements. Yet during the night the pendulum had moved insidiously around, smashing aside the lines of trestles he had placed along its original route as safety barriers, and as he admired his handiwork—"

"The pendulum struck him on the back of the head, killing him instantly, while he stood in a spot he thought safe!" I exclaimed. "But then as the hours passed, the pendulum's line of swing continued to turn, and by the time we arrived, it was far from the site of the body, so we assumed it could not be implicated."

Summerlee shook his head sadly. "I am sure the accident happened because his Lordship, although in his way an intelligent man, half believed his own notions about the validity of old beliefs," he said. "But while linear motion is perhaps a relative matter"—I remembered the example of the trains—"rotation is an absolute quantity. Even if we lived in deep caves, and had never seen the stars, nor had any idea of the existence of the Solar System, we could still tell in half a dozen ways that the Earth turns on its axis, and at what speed."

"Not merely by the behavior of pendulums?" I asked.

"Hardly! Take the gyroscope, for instance: a flywheel mounted on gimbals will keep pointing in the same direction, irrespective of the motion of its mount. But the behavior of a pendulum is easier to comprehend: the forces on a gyroscope are difficult to measure in practice, and quite subtle to calculate mathematically, as a result of which much nonsense has sometimes been written about these devices.

"Rotation also gives rise to electrical effects: the turning of an electrically charged body generates a detectable magnetic field, for instance. And I could give further examples. Only an ignoramus, sir, could nowadays have any truck with the concept of a non-rotating Earth."

As the four of us walked to Vauxhall police station to report the tragic accident, I was tempted to challenge his arrogance.

"Granted that the Earth rotates," I asked cautiously, "is it not just possible that it is otherwise stationary in space, and the Sun and planets do in fact revolve about it in epicycles? Surely we could never know the difference."

"We can detect the Earth's motion directly with respect to the nearby stars, whose apparent positions in the sky shift by an amount just great enough to be visible by telescope as the Earth goes around its orbit," replied Summerlee contemptuously. "But even without these measurements, I should still have no hesitation in rejecting epicycles, for there is no underlying pattern to them. An arbitrary set

must be constructed for each planet. If a new body came along—
say, a comet, as happens from time to time—there would be no way
initially to use epicycles to predict its motion. Whereas understanding that the new body is in reality in solar orbit under the Sun's
gravitational pull, such a forecast can readily be made.

"The trouble with epicycles is that by taking a sufficiently great
number, a set can be contrived to describe any arbitrary motion. If I
gave you a graph of a drunk's meanderings along Piccadilly, you
could fit the trajectory with epicycles. But your trouble would not
tell you anything useful. The principle of Ockham's Razor, sir, is
essential to science: always adopt the simplest hypothesis, which
requires the least assumptions, capable of describing the known
facts. Without that guiding principle, we could fill our heads with all
sorts of wild and fanciful notions which could never be proved or
disproved, using mental capacity of which some of us, I fear, have
little to spare." He looked around at us disdainfully, and took his
leave.

"Well, I have learned something today, Watson," remarked
Holmes as we strolled back toward Baker Street in the bright afternoon sun.

"That the Earth rotates?"

"No, a little more general than that. Not long after we first met, I
teased you that I had not even heard of the Copernican theory."

"I recall it well: I wrote about it in *A Study in Scarlet*."

"And in truth, I considered it a fact of no possible relevance to
me, that the Earth goes around the Sun rather than vice versa. Yet
today my ignorance could have caused an innocent man to be sent
to the gallows. From now on, Watson, my mind shall be a little more
open to matters scientific."

2

The Case of the Missing Energy

"SOMETHING OF A MYSTERY, THIS gentleman," commented Sherlock Holmes as we heard the tread of our morning's client upon the stairs. "When he spoke to Mrs. Hudson in our absence yesterday, he was apparently willing to divulge his name, which is Morrison, and nothing else at all."

Morrison was ushered in a moment later, and I looked at him keenly. I have often admired my friend's ability to make deductions from specific detail, but it seems to me there is something to be said for forming a general, intuitive impression: the holistic approach, as a diagnostician would call it.

Yet I was forced to admit no very useful insight emerged. His sunburned face and calloused hands suggested a laborer, whereas the smart suit and pocket fountain pen clearly indicated a professional man. He moved with an odd, slightly rolling gait, and although he inspected us both with a clear and steady gaze, one hand trembled oddly.

"Delighted to make your acquaintance," said Holmes, waving our visitor toward a fireside chair. "I trust your sea voyage was a pleasant one, and the diving work undertaken with due caution."

If Holmes's intention was to put our visitor at ease, he was hardly successful. Morrison sprang to his feet with ashen face.

"Who told you of our expedition?" he shouted. "We have been betrayed, as I suspected. I demand, tell me your source!"

Holmes leaned forward, much interested. "Sit down, sir, I beg you. No one has told me anything about you. Your gait told me that you have not been long ashore and still have your sea legs, and your tremor is a classic sign of what my colleague Doctor Watson would diagnose as caisson sickness: you have at some recent point been diving, and then decompressed too suddenly."

Our visitor relaxed visibly. "Forgive my nerves," he said. "I am, as you deduced, a marine engineer, and normally I would have no objection to the whole world knowing of my business. But I have just been on a sea trip to a location of the utmost secrecy, working for a most demanding man, who from his manner would be quite capable of flaying me alive if I or my crew were guilty of indiscretion. Although he acts like a madman, he is convinced that he is on the track of vast riches and that secrecy is vital. And perhaps he is right after all, for if we are on a wild goose chase, who would bother to sabotage us? But there again, what has befallen us seems almost beyond a mere human agency: half the time I am convinced my superstitious crew are right, and the triangle of water within which we work is indeed accursed."

With that he shut his mouth firmly. It seemed to me he had given away more than he intended. Where could he be referring to but the mysterious Bermuda triangle so feared by mariners, and what could merit such an expedition but sunken treasure? I could see the Spanish galleon in my mind's eye, coffers of gold lying waiting on the tropical sea bed.

Sherlock Holmes smiled. "I can assure you that, in a career filled with mysterious happenings, I have yet to meet my first ghost or supernatural manifestation of any kind.

"Now, normally I make it a rule never to tolerate mystery at both ends of a case: frank disclosure by the client is a first requirement. But in the present instance, I can see why secrecy as to your destination is vital. Tell me only what mishap has occurred, and as much of the circumstances as you see fit."

Morrison nodded. "I will do so. The start of the affair will sound mundane to you. Our firm was approached by a well-known scientist—you would probably recognize his name—who had formed a theory about valuable deposits of a certain mineral on the sea bed. He wished to charter a vessel equipped with a diving bell, and certain more specialized equipment, to test out the idea. Such an expedition is not a cheap matter, but he had ample funds: evidently he had convinced some wealthy backers.

"The request would have been routine but for two things. The first was the nature of the mineral he expected to find. The second was the personality of the Professor himself. A more aggressive, arrogant, and intolerant man you could not hope to meet. But he is by no means stupid. I have met scientists whose grasp of practical engineering, or indeed anything else outside their specialist field, was sadly lacking, but not only does the Professor grasp details quickly, he is soon telling you with great confidence how to do your own job. Truly an infuriating man."

He collected himself somewhat. "At all events, we set sail from Lowestoft, and by good fortune the Professor was detained on other business. Really, I think I should have strangled the man if forced to endure his companionship aboard a small vessel! We followed our directions to the letter, but as we approached our destination, odd things started to occur. Most could have been explained by the mishaps and coincidences which do happen at sea, but one occurrence was bizarre. As we steamed past the mouth of an estuary, the ship came to almost a full stop in the water. There was no trace of a current, no breath of wind, the propellers churned at full speed, yet it was as if the *Matilda Briggs*—" He bit his tongue.

"You may rely absolutely on my discretion, and on that of Doctor Watson," said Sherlock Holmes soothingly.

Our visitor hesitated for some seconds, then seemed to come to a decision, and sat back with a more relaxed air.

"I suppose that I must place my trust in you. Well, it was as if we were steaming through glue. The *Matilda Briggs* is a powerful vessel, capable of a full ten knots, yet with the engines turning at maximum revolutions she was all but stationary in the water. You could almost have believed some great invisible beast was dragging us backward.

"At any rate, we ploughed on and in due course reached the designated spot. We prepared the bell for the dive. Are you familiar with the disposition of a diving bell?"

We both nodded, but Morrison drew a sheet of paper to him, and with the clear, economical strokes of a practiced draftsman drew the diagram shown on the opposite page.

"The device is simplicity itself. The bell is suspended from a winch, and is open to the sea at the bottom. A powerful pump on the ship thrusts air down a connecting tube. The air pressure is sufficient to keep the bell from flooding, just as you can plunge an upturned tumbler into a sink without the interior filling, and in fact surplus air continually bubbles from around the bottom rim: the turnover must be sufficient to keep carbon dioxide from accumulating, so the air remains pure to breathe. And that is all there is to the thing. There is no kind of machinery or fuel—indeed, nothing flammable or electrical—on board the bell, and no connection to it other than the hoist cable and air tube."

"Then there is no means of communication between it and the ship?" asked Holmes carefully.

"That is correct. Otherwise, there would be no mystery to consult you about." Morrison took a deep breath. "Before we undertook a live dive, we performed the customary unmanned test: we started the pump and winch and lowered the bell to just above the sea bed. We were able to hoist it back without difficulty, and the interior remained dry as the Sahara.

"Then two of our most experienced divers took their places in the bell, and it was lowered again. We were to give them just an hour on the sea bed, but the trip took much longer, for we had to pause every few meters to allow them to acclimatize to the pressure change."

"And you are sure the intervals allowed were sufficient?" I asked.

"Certainly. The dive was not to any exceptional depth, and both men had undergone identical trips before without any harm

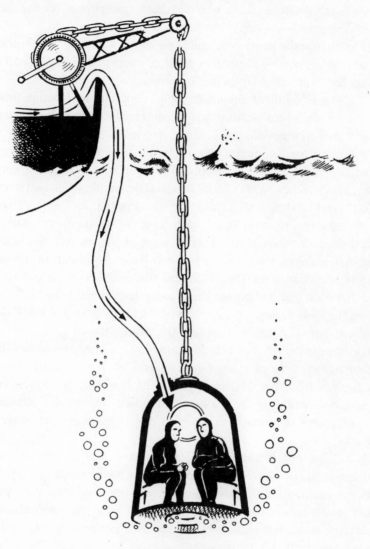

The Diving Bell

coming to them. So there was nothing to prepare us for the sight which greeted us when the bell was swung back aboard."

Our guest's hand was trembling more violently than before, and Holmes poured a splash of whiskey into a glass and passed it to him. He took only a sip before continuing.

"The equipment aboard the ship, winch and air-pump, functioned perfectly throughout the dive, and the bell was pulled clear of the water on schedule and lowered into its cradle on the foredeck. We waited for the men to scramble out beneath—one is always keen to escape the claustrophobic confines, even after a short period—but there was no sign. I ducked beneath the bell and stood, and beheld a sight which made me doubt my sanity.

"The two men lay dead upon their bunks, their eyes staring, their skin strangely mottled. They were almost naked; the heavy garments divers always wear to keep warm had been discarded. And the cork matting which lines the bunks had been hacked away so that each man lay pressed against the bare metal beneath."

He took a gulp of the whiskey. "I know you will think this foolish, Mr. Holmes, but among the Scandinavian crew I heard murmurings from which I could distinguish the dread word *Kraken*. That is the name of a legendary sea monster of colossal size— it derives from the Midgard serpent of Norse myth, which encircled the world on the sea bed, its tail in its mouth—which supposedly emerges now and again to drag hapless sailors to their deaths.

"Of course, I sent ashore for a doctor, and he was able to diagnose the cause of death with certainty. It was heatstroke. Heatstroke, surrounded by sea-water nearly at freezing point, and with no possible source of heat present! Really, a sea serpent would have been more believable: I am sure there are stranger creatures in the ocean depths than men ever get to see.

"It seemed to me there were but two possibilities. One was that the laws of physics, at least as currently understood, were being violated. A truly bizarre theory occurred to me. You have heard of the phlogiston theory of heat, which was widely believed until the middle of this century?"

I shook my head.

"The concept was that heat was a kind of invisible gas which permeated all substances. In its free form, it could seep both within a substance, and between any two items placed in contact, seeking to equalize its pressure as a gas does. This explains why a hot object warms a cold one. There was also a bound, locked-up form of phlogiston which any fuel or burnable substance contained: this latent phlogiston was released by flame, in the process of combustion."

"It seems quite a consistent theory to me. Perhaps it is an equally valid way of looking at things as the more modern notion of energy," I commented.

"Well, my father and grandfather would have agreed with you; but modern savants have rejected the concept. Their disproof is that a turning shaft, whose end grinds against a whetstone, can apparently create new phlogiston by friction without limit, just as long as the motion is continued. Similarly, repeated mechanical hammering can heat an object indefinitely. A distinct paradox, if you believe in phlogiston as an immutable substance which can be neither created nor destroyed.

"So, the concept of phlogiston has been replaced by the more general notion of conservation of energy in all its different forms. There is kinetic energy, the energy an object in motion possesses; there is potential energy, which an object in a high place loses as it descends; and there is the thermal energy possessed by a hot object. Some suggest thermal energy may be a form of kinetic energy, caused by the atoms of which all matter may be composed rushing about in ceaseless motion. But I am a hard-headed engineer, and will believe in atoms when I see one.

"Now just suppose, gentlemen—I know it sounds fantastic—suppose phlogiston is, after all, real. You recall that for a time our ship was almost at a full stop in the water, for all that her propellers were churning away, as we approached our diving spot. Where did the energy of the propellers then go?"

"Why, into a cloud of phlogiston!" I cried. Then I was struck by an even more ingenious insight. "And that very cloud of phlogiston might have floated beneath the water, until it came in contact with your diving bell, heating it to such a temperature that although your divers stripped and pressed themselves against cold metal in contact with the outside water, they were overwhelmed and succumbed."

Our visitor nodded, but Holmes appeared none too pleased at the idea.

"Frankly, sir," he said, "my specialty is investigating breaches of the laws of man, rather than of the laws of physics. And fortunately for my trade, the laws of man are only too often violated, in contrast to those of physics. You spoke of a second possibility which had occurred to you?"

"The second possibility was that some fiendishly cunning saboteur was at work. So I returned to London, determined to obtain the most expert help with the problem, whatever its nature," Morrison continued.

"I first sought our sponsor, who I may as well tell you is Professor Challenger of Imperial College, London. For all his eccentricity, he had impressed me as a most intelligent man. But he was out of the country on some mission. So I turned to the Royal Society, that official body comprising the nation's most eminent savants. I would have thought any scientific mystery attested by reliable observers would be of the greatest interest to them.

"Alas, I was disappointed. I did try to describe my phlogiston theory to one man who would listen to my story, but he dismissed it with contempt: he evidently thought the evidence against phlogiston so convincing that only an ignoramus or a crank would consider the notion."

Sherlock Holmes smiled. "That is ironic. In 1847 one James Prescott Joule approached the Royal Society with a detailed set of experiments disproving the existence of phlogiston. In addition to mechanical work, he showed that electrical energy, or even the potential energy of a falling object, could be converted to heat."

"How could he have done the latter?" I interjected skeptically.

"He measured the temperature of water in a river at the head and foot of a waterfall, and found the fall had raised its temperature. But notwithstanding his ingenious demonstrations, the Royal Society rejected his paper, claiming the whole matter of phlogiston was already too well understood to need further investigation. It was only when he started to broadcast his ideas to the public, starting with a famous lecture at Saint Anne's church in Manchester, that they were forced to take notice. It was intelligent laymen who were first convinced by his logic, rather than the scientific establishment.

"Perhaps it demonstrates that in matters of science we laymen should still have a role to play, and that ordinary men of common sense can sometimes be wiser than savants gifted in mathematics. But we digress. Pray continue your very interesting story."

"Well, Mr. Holmes, the only man at the Royal Society who was prepared to listen to me at all was one Professor Summerlee. He says that he will in due course come to investigate. But it is clear he thinks I must be either a fool or a liar, and is coming only in the hope of disproving my version of events."

"We have met Summerlee, the notorious skeptic," said Holmes. "And I have heard of Challenger, whose utter inability to use tact or restraint when debating his more outrageous notions has almost made London too hot to hold him: no wonder he has taken a break abroad. But let us move on from these scientific matters. Presumably, you are seeking my help with the second possibility?"

"Exactly so. If there is no scientific mystery, then sabotage is the only explanation. If so, it must have been done with extraordinary imagination and cunning. And all advised me that you were the man to investigate such a crime."

Sherlock Holmes nodded at the compliment. "I should be delighted to do so. The snag is that pressing business in London precludes my making any voyages just at present. However, I feel I will be able to work matters out from here if my coinvestigator is able to visit the site. How would you feel about a little sea-trip, Watson?"

I sprang to my feet. "I am your man," I declared warmly. "My childhood dreams may have taken some time to realize, but a voyage to the Bermuda Triangle in search of gold on the seabed! Have no fear, I shall be delighted to come."

Our visitor looked blank. "Gold? Bermuda Triangle? It is of the

Barents Triangle I have been speaking, Doctor, a bleak region just off the Norwegian coast. And it is not gold we seek, but oil—or black gold, as some have come to call it. But of course you are jesting. The *Isis* sails from Lowestoft tomorrow afternoon. You will be aboard? Capital!"

As soon as our visitor had left, I felt some misgivings descend. "You know, Holmes, how I am delighted to help out, but in the past when you have sent me to investigate on your behalf, I have not invariably been successful."

"Well, in fact, Watson, I was less than candid with our visitor. I do not believe his sabotage notion for one moment. Professor Challenger's theory that there are vast reserves of oil beneath the North Sea is viewed by others as the raving of a crank. Not worth investigating, never mind going to the lengths of deadly sabotage. I think we are dealing with some simple oversight or natural phenomenon."

"But Holmes, that makes it worse! If learned scientists and experienced engineers are baffled, the matter will certainly be beyond me."

"On the contrary, Watson, I consider you ideal for the task. Ingenious men are precisely the most naive and gullible when it comes to spotting the obvious. Let me tell you a story from a few years back. A famous scientist came to me for advice. He had been offered the chance to purchase a machine for turning lead into gold.

"Now you or I could have told him that this was impossible. The most basic law of chemistry is that one elemental substance cannot be turned into another. There are less than a hundred such elements, and the quantity of each is always perfectly conserved, and no doubt has been for all time. Neither gold nor any other element can be created, destroyed, or transmuted by any known means.

"But this man was the victim of his own intelligence. He listened to lengthy and completely spurious accounts by this modern-day alchemist of how his machine supposedly worked. He was fooled by a little sleight of hand involving a lead brick coated on three sides with gold leaf, which was substituted for one of genuine gold when the machine was operated. Any amateur conjurer could have spotted the trick at once; yet one of the most eminent minds of our age was taken in.

"There are problems that call for imagination and ingenuity, and there are others that call for a total lack of it. A dogged reliability and honesty can be far more valuable. I cannot imagine a man better fitted to this task than yourself, Watson."

So it was that on the morrow I journeyed to Lowestoft and boarded the *Isis*, a small tramp steamer which would take me across the North Sea to the *Matilda Briggs*'s mooring, a voyage of some three days. I have never been a good sailor, and kept to my bunk for the first part of the journey. My incentive to rise was not improved by what I heard of my one fellow passenger. He had joined the ship just as we sailed, somehow browbeating the mate into letting him board. Periodically, his bull-like voice echoed through the bulkheads as he rebuked various members of the crew, frequently demanding to see the Captain, and I felt little desire to make his acquaintance.

On the second day, however, the sea conditions improved considerably, and in the evening I took myself up on deck. My co-voyager was propped against the rail, great shaggy head nodding as he seemed to half-doze. But as I stood beside him, a sardonic eye turned on me.

"Gazing, like Odysseus, upon the wine-dark sea?" I ventured rather fatuously.

He raised an eyebrow. "Indeed, sir, they say that the Universe is in a glass of wine, but you will observe more of its nature in a somewhat larger context."

He waved an expansive hand. "Take the horizon before us. The ancient Norsemen considered the world was flat. Yet if you look at the horizon, as they must often have done, do you not clearly see the curvature of the Earth betrayed?"

I looked. It might be that from the high deck of a liner the curvature would be visible. But we were only a few feet above the waves, just as the Viking longboatmen must have been, and even in a calm sea, the precise shape of the horizon was not discernible. In any case, the eye is poor at judging gentle curves: the columns of Greek architecture must actually bulge slightly, in order that the edges will appear straight when seen from beneath. I said as much to my companion.

To my surprise, he nodded in approval. "Quite right, sir! You are capable of observing what is to be seen, rather than what it is suggested to you should be seen."

He ruminated. "Still, a thinking Viking might have been able to guess. Did they think the Earth's surface went on forever?"

"I think not; it would scarcely seen credible."

"But would you really believe in an edge to the world, with the sea pouring off continuously at such a rate it would be gone in days?"

"When you put it like that, it hardly seems likely even primitives could really have credited it."

"And supposing you were to confront an ancient Norseman, and tell him that the flat surface on which he stood neither had an edge, nor went on forever, what would he think?"

"Why, that it was a paradox!" I cried. "A plane surface that neither has an edge nor goes on forever is simply a contradiction."

"Exactly! And the mere consideration of the paradox, without the need to go making observations of the horizon, or any other external clue, might have led him to the conclusion that since there are no real paradoxes, some assumption must be wrong. And there: the flat-Earth hypothesis is demolished. The Earth must curve back into itself, he would realize, and implausible infinities and boundaries vanish.

"And what," he said suddenly, "do you make of the paradox we are going to investigate?"

I looked at him keenly, and remembered my discretion.

"It is all right," he said. "I am Professor Challenger, initiator of the expedition, and you may speak freely to me. I returned to England yesterday by chance, and came at once. Absurd to think that a mere detective should be required, when George Edward Challenger is on the job! But let us hear your conclusions."

"I would call them guesses at the moment," I said. "But Morrison's idea of a phlogiston cloud—you have heard it?—has at least the merit that there would then be one mystery rather than two."

He snorted. "Truly a sign of the undisciplined thinker, to assume two events must be connected merely because they are both unexplained."

"You think the scientific establishment's rejection of the phlogiston theory definitive, then?"

He seemed to swell. "Never assume anything merely because most supposedly wise men believe it!" he said. "The whole history of scientific advance is that only when a pack of learned fools are shown conclusively to be in the wrong is progress made."

"So it is rather the iconoclast we should believe?" I sought to conciliate him.

"Certainly not. In the vast majority of cases, those with wild new theories are completely wrong. The point is that you cannot judge a theory by the person who proposes it."

"Well, I would be more inclined to believe a theory proposed by someone who already has a track record of proven success in the world of science," I said.

"And so would many. But you would still be wrong. Take Sir Isaac Newton's mystical beliefs. The greatest scientists have mixed insight amounting to genius with the most absurd follies at other times. Kindly biographers tend to overlook these blunders."

"So, only the most basic qualities such as honesty should be taken into account when first evaluating a man's theories?" I asked.

"That would seem logical. But in fact we know that many great scientists cheated to generate the data that validated their ideas. Take Mendel's work on plant breeding: his data are almost as implausible as if a statistician were to claim that each time he tossed a penny a thousand times it fell exactly five hundred times heads and five hundred times tails. Yet his ideas were correct for all that. The concept I am trying to get through your head, sir—with remarkable difficulty, I might add—is that you must think about the idea itself, not the proposer."

"Then new experiments must be devised to test each new idea?" I asked.

"If necessary, but, above all, the matter must first be properly thought through. Before acting, consider whether the idea is consistent with what is already known, and what its consequences should be. A thousand ill-thought experiments may reveal nothing you did not already know," said Challenger firmly.

"Still, often experiment is needed," I commented. "For example, before there was machinery which could turn a shaft rapidly and continuously, there can have been no observations of heat generated indefinitely at a point, the experiment which overthrew phlogiston theory."

Challenger looked at me with undisguised scorn. "Really? Do you know how primitive man generated fire? He sharpened a stick, placed it against a piece of bark, and twirled it rapidly until the heat produced was sufficient to ignite it. A technique within the grasp of our ape-like ancestors, but quite beyond the luminaries of our Royal Society!

"No, sir, just as with the matter of the shape of the Earth we were discussing, just now, it was not new experiment, or additional observation, that was required: it was the contemplation of a paradox. You posit phlogiston, a substance insubstantial, but nevertheless a real element whose amount is always conserved. You note the commonplace that in fact a turning shaft or beating hammer can produce phlogiston indefinitely. Lo!—a paradox. And since reality can contain no paradoxes, we must rather conclude no phlogiston."

"At any rate," I said, "I can understand why the whole phlogiston business would have been thought less than important in practical terms. More a question of scientific theology."

It appeared that in my attempt to calm the conversation, I had finally overstepped the bounds of his self-control. "Really!" he cried. "You have the good fortune to live at the first time in history when most men are freed from heavy labor, to engage in more pleasurable and profitable pursuits. Freed from toil by the steam engine, whose perfection has enabled the locomotive, the traction engine, and now the electrical generating plant. In a hundred years men may take such things for granted, but you, sir, should remember that only a proper understanding of the laws of energy conversion by men of science has lately relieved you from that burden. Ignorance and ingratitude, such is the lot of the savant. Good night!" With great suddenness, and to my considerable relief, he turned on his heel and left me.

THE FOLLOWING DAY, I arose quite free of mal-de-mer, and was able to eat a full breakfast in the lounge in the more pleasant company of the ship's captain. In fact, so heartily did I eat that I felt a few brisk turns around the deck would be necessary to maintain a healthy balance—as a scientist would put it, to dissipate by exertion the chemical energy of the food I had eaten.

On emerging, I perceived my twofold mistake. First, the wind, although not strong, was exceedingly cold, a fact unsurprising given that the bleak, snow-covered Norwegian coast was already looming to starboard. Second, my disputatious companion of yesterday had evidently had the same notion.

"Good morning," he greeted me with unexpected warmth. "A splendid day, is it not? I have been thinking of the problem that awaits us, and contemplating the clues about. Truly, a ship at sea is surrounded by energy, is it not?"

I gazed about in some bewilderment. The surroundings seemed devoid of energy—certainly of the heat kind—to my perceptions.

"Take the ship we are aboard. Steaming as she is at some ten knots—near enough, five meters per second—she contains quite a quantity of Joules."

I recognized the name of the amateur scientist who had disproved phlogiston to the Royal Society, but must have shown my bafflement.

Challenger snorted. "I am referring to the units named after James Prescott Joule," he said. "After his death, the scientific community made amends for their initial disrespect by naming the metric unit of energy after him. The rate of consumption of energy was named after another James: Watt, of steam engine fame. A device which consumes one Joule of energy per second is thus using one Watt."

"I really preferred the old units," I said with nostalgia.

"Well, they are easily translated. For example, one horsepower is roughly seven hundred and fifty Watts, so a four-horse carriage is drawn with a power of three thousand Watts, usually called three Kilowatts."

"I still have no very clear picture of what a Joule is."

"Well, nowadays we think of energy as force multiplied by the distance through which it acts. To use familiar examples, the energy required to raise a one-kilogram weight a distance of one meter is about ten Joules. But in terms of heat, to warm the same kilogram of water by one degree Centigrade consumes some four thousand two hundred Joules. The energy required to raise water from freezing to boiling point is therefore equal to that needed to raise it vertically some forty-two kilometers, or twenty-six miles.

"Now, I was idly considering how much kinetic energy— energy of motion—the ship we are aboard currently possesses. Take her weight as one hundred tonnes, one hundred thousand kilograms, and her speed as five meters per second, and what do we have?"

"I suppose that you simply multiply the speed by the mass, to give half a million Joules?"

The Professor looked at me with contempt. "No indeed. The momentum of an object—the amount of its inertial motion in a given direction—is indeed a quantity conserved in physics. But the energy is proportional not to the speed but to the square of the speed. For example, if we were traveling at twenty knots rather than ten, then—"

"We should certainly be in contention for the Atlantic Ribbon as the fastest steamship afloat," I interjected facetiously.

"Our energy of motion would be not doubled but quadrupled," he finished.

"Now there," I observed, "is a definition which seems to me quite arbitrary. Why could you not have an equally valid system of physics in which the energy is directly proportional to the speed?"

"Because the truth would emerge as soon as you converted the energy into a different form. If you brake an object mechanically to rest, for example, frictional heat is generated: and an object moving twice as fast will produce not twice, but four times as much heat."

"Well, careful experiments must certainly have been needed to confirm this relationship," I said skeptically.

"No, sir, once again the mere contemplation of paradox would suffice. Consider, for example, a falling object. When you raise an object, the work done is obviously directly proportional to the height through which it is raised."

"And to its weight, surely?"

"Quite so: distance multiplied by force. But now having raised the object, let us drop it. The energy of motion it acquires in falling a given distance is obviously the same as the energy required to lift it through that distance."

"That seems clear enough."

"Now, the Earth's gravity accelerates any freely falling object, adding ten meters per second of speed in every second it falls. After one second, it will be moving at ten meters per second, after three, at thirty meters per second, and so on."

"Doubtless explaining why a fall from any height is medically inadvisable," I said.

Challenger ignored me. "So after one second, how far will an object have fallen?" he asked.

"Well, at the start of the second it is stationary, and at its conclusion it is moving at ten meters per second. So the average speed is five meters per second, and it will have fallen five meters. Sixteen feet—why, that is correct!" I said with some surprise, re-membering my schoolroom teaching.

"And after two seconds?"

"Well, the top speed will then be twenty meters per second. An average of ten meters per second, for two seconds: twenty meters. Why, the distance fallen increases simply as the square of the time, does it not?"

"It does. Now in this second case, the object has fallen twenty meters compared with the first object's five meters, so the energy conferred on it by gravity must be four times greater. Yet its speed—"

"Is only twice as great. So the energy is indeed proportional to the square of the speed!" I said.

Challenger wagged a finger at me. "Quite so. And you, the lazy but wise savant, have discovered the fact without the necessity of moving from your armchair."

Considering that we had both been strolling briskly along the deck throughout the conversation, I thought this a slight exaggeration.

"A little more contemplation," Challenger continued, "would lead you to the formula that the kinetic energy of any object is exactly half the mass multiplied by the square of the speed. And unlike the pedantic experimenter, whose measurements can never be perfectly exact, you know that your formula must be *precisely* true. Otherwise, you could construct some system whereby an object was lifted against the force of gravity, dropped, and its energy of motion recovered, which would yield—"

"A perpetual motion machine!" I exclaimed.

"A thing which all experience would indicate is paradoxical. The quantity of energy in the world is always perfectly conserved, and no exception has ever been found. If it were, the Universe would be a rather different place."

"The price of coal shares would fall sharply, for one thing," I commented light-heartedly. Then I remembered that if Challenger's theories of oil secreted beneath this inhospitable sea were correct, a fall in coal prices was entirely possible.

"So, as I was about to say before your little digression," Challenger continued, ignoring my raised eyebrows, "the total energy of motion of our ship works out to be just over one million Joules: in terms of heat, barely enough to boil a kettle. Certainly not enough to explain the divers' bizarre deaths!" He raised his finger. "What other sources of energy do you see about us?" he asked.

I looked at the low swell on the sea which was outracing us toward land. "Well, the motion of the sea as waves, for one thing. The water in those wave tips looks to be moving faster than us, and must certainly hold some energy."

Challenger shook his head. "You should not confuse the illusion of motion with its actuality. No particles of water are actually traveling with the waves you see. In fact, each individual drop of water is describing a circular motion which takes it nowhere overall. You can demonstrate the fact by dropping a cork of oak wood, which has almost the same density as sea-water, overboard, and see it cycling in place beneath the surface. Many phenomena mimic the effect of motion. I have seen a row of chorus girls at the opera stand at the front of the stage and simulate waves by raising and lowering blue-painted boards in synchronized time. The waves sped along the footlights, but the motion of each individual board was much gentler, and no actress actually moved from her position."

"Then a wave is mere illusion?" I inquired.

"Perhaps it is more a matter for philosophers than scientists, but I would say not. A wave is perhaps a *process* rather than an *entity*, but nonetheless real for that. A wave certainly contains energy: potential energy, as the wave tips are raised above the mean sea level, and kinetic energy, of the rotating particles of water."

"I see. But of course the wave cannot actually transport energy, as the water particles do not move along." I said.

"That does not follow. Forces in motion transmit energy, even though no mass is transported. In fact, I once proposed to Her Majesty's Government that a system of floats and pulleys deployed along the west coast of Scotland could harness the energy of the Atlantic waves rolling in to generate almost limitless amounts of energy. But I was laughed at for my pains. Galileo, Joule, Challenger—the paths of us pioneers of science are indeed roads to martyrdom."

I had difficulty keeping a straight face at the man's conceit.

"But let us turn to the problems ahead of us," he said. "We are faced with one phenomenon which subtracts energy, and another which adds it. I would be glad to hear your thoughts."

I was flattered that despite the man's gruff manner, he was taking me seriously.

"I have seen that where paradoxes, or inexplicable events, seem to occur, then it is one's initial assumptions that must be questioned," I said carefully. Challenger nodded encouragingly.

"Well, you and other men of science seem to assume two things which I find rather implausible. The first is that the laws of Nature are capable of *exactly* conserving many different quantities. If human measurement is never perfectly accurate, always admitting of accumulating errors, can Nature really be so perfect?

"The second is that the laws of physics must be exactly the same everywhere. Perhaps they are quite different on Mars or Jupiter or the far stars. Indeed, why should they be identical everywhere on Earth?" I looked at the bleak, eerie landscape nearing us, so utterly different from the comfortable London world of human construction.

Challenger raised his massive eyebrows tolerantly.

"I suppose that somewhere, Doctor, there might be universes as you describe, where the laws of Nature vary from place to place, and where randomness somehow intrudes to make such laws as there are inexact. What hells they must be for any scientists that inhabit them—if intelligence could evolve in such inconsistent places, which I doubt.

"All I can say, Doctor, is that these properties of our Universe have been better tested than you can probably imagine.

"First, the Universe appears perfectly symmetrical in the ways it enforces its laws. They are the same however far you travel in any direction in space. Otherwise the measurements of the properties and motions of distant stars we make with telescopes and spectrographs would not be as they are. The laws are also the same with respect to displacement in time, or our Solar System and the environment of this Earth we stand on could hardly have remained stable through billions of years of evolution. And they are the same irrespective of orientation, or various optical and mechanical devices, especially rotating ones, would perform imperfectly.

"Second, certain quantities are conserved under all circumstances that we know of. Of course, you must define these quantities carefully. To be frivolous, laps are not conserved, for where does your lap go when you stand up? And phlogiston also fails the test. But certain basic things—mass, energy, momentum, electric charge—are preserved to an amazing degree of precision, to one part in billions of billions of billions, or we should easily know it.

"Come to think of it, the number of conserved quantities we know of is similar to the number of symmetries the Universe is observed to possess. I wonder if there might be some deep link here." He fell silent, musing.

I sought to be helpful. "I suppose that more and more conserved quantities will be discovered as science advances? Electric charge, for instance, is a comparatively recent discovery."

Challenger tossed his head in annoyance, as if I had disturbed some delicate but promising train of thought. "If anything, the opposite. Construction of an appropriate machine shows that phlogiston is but one form of that more general thing, energy. Perhaps more ingenious experiments may yet show that quantities we consider to be quite different are merely alternative manifestations of the same essence.

"The more limited your technology, the more numerous the apparently conserved essences you perceive. For example, in a world where the only available experimental devices were slow-acting pulleys and levers, you might think that gravitational potential energy was conserved: it would be possible to raise a one-kilogram weight two meters only by allowing a two-kilogram weight to fall one meter, and so on. Yet in reality the energy could be converted to and from quite different forms, such as heat.

"But, Doctor, let us return to the problem at hand. A ship that loses energy, a diving bell that gains it. I have a few thoughts. Let me give you some clues." He pointed to the sky. "Do you see anything mildly unusual about these clouds?"

I looked up. There was indeed something curious about them: two sets of otherwise similar clouds were traveling in different directions, seemingly crossing over one another.

"Ah, the wind is traveling at different rates and directions at different heights," I said.

The Professor nodded. "And does that suggest anything to you?"

I pondered. "I remember a story of a tall-masted ship which sailed past a lesser vessel completely becalmed. The topsails of the former were in a brisk breeze, even though there was a flat calm at sea level."

Challenger nodded encouragingly. "But, Professor," I protested, "the *Matilda Briggs* is a steam boat, with neither masts nor sails."

Challenger shook his head in wonderment. "I am really beholden to you, Doctor," he said. "I have at times been criticized for a certain impatience toward my students, who usually seem to me a singularly dull-witted lot. Indeed, I am rarely asked to undertake teaching duties of late. But you have opened my eyes to the true range of variation of the species to which I belong. Henceforth I shall certainly be more tolerant!"

While I pondered this enigmatic compliment, Challenger spoke in a slow, patient tone: "I am thinking not of the motion of the wind at different heights but of the sea at different depths. The tides around the Norwegian coast produce strange currents—you have heard of the dreaded Maelstrom whirlpool?—and it is known that tidal currents can vary rapidly with depth. A strong submerged current could act upon the keel of a ship, dragging her back, even though the water at the surface was stationary. The strongest known tidal current in these parts—through the Pentland Firth, which separates the tip of Scotland from the Shetland Islands—reaches sixteen knots at the spring tides, easily strong enough to produce the required effect."

An idea seemed to occur to him. "Doctor, you would be the ideal person to assist me in a small experiment. Could you come to my cabin in a few minutes?"

On entering his cabin, I saw that he had arranged a paraffin lamp somewhat precariously upon the side table. A glass aquarium tank, containing water but at present empty of fish—no doubt ready for specimens he hoped to acquire on our voyage—stood beside it, as did a pair of bellows. To my surprise, he locked the door behind me.

"Now, Doctor," he said, "if you were embarking on a journey of several hours in an uninsulated cast-iron vessel surrounded by freezing sea, would you be tempted to take some source of heat along—a paraffin heater, perhaps?"

"Morrison told me they have a strict law aboard ship against such actions, rigorously enforced by the Captain."

"Well, the Captain may be second only to God in his absolute command, but given a choice between the laws of men and the laws of physics, I think I know which are the more easily broken. A small heater could easily have been smuggled into the bell, and lit once it was beneath the water. I shall now show you what may have befallen the perpetrators."

He lit the paraffin lantern and raised the bellows. "This lamp is designed to burn safely. But the greater the supply of oxygen, the fiercer the flame. You recall that air was continuously being driven down to the bell by a powerful pump."

He worked the bellows, and the lantern burned more brightly.

"I see no catastrophe yet. They would surely have been careful about it," I said.

"Ah, but as the bell sank deeper, the pressure of the air in her grew. Remember, she is open to the sea at the base. Ten meters down, the air density inside will be twice that at sea level, and the partial pressure of oxygen correspondingly raised. Any flame will burn faster."

He worked the bellows furiously, with disastrous results. The flame of the wick was blown down to touch the paraffin in the body of the lamp. A moment later, a fierce meter-high flame had arisen.

I thought quickly. There was ample water in the aquarium beside the lamp, but it was too heavy to lift. There was a parable about Mahomet and the mountain. I lifted the lamp and tossed its contents into the glass tank. But it was the worst thing I could have done. I should have remembered that paraffin and water are immiscible, and paraffin is the lighter of the two. The paraffin spread over the water, and in a moment the whole surface of the tank was ablaze. Luckily my companion reacted with great speed, grabbing a piece of sacking and laying it over the top of the tank. In a few seconds, deprived of fresh air, the flames had died.

"Well, let that be a warning to you against foolishly designed experiments!" I cried.

"Not at all, Doctor. You have just helped prove my theory most beautifully. I simulated the confines of the bell by locking the door here. The fire burns out of control in the copious oxygen. No escape, and seconds to react. What do the men do? They tip the burning heater into the sea at the base of the bell. But the paraffin floats out, the greater surface area hastening the burning. The heat grows rapidly; the fumes no doubt hinder any further action. The men die. By the time the bell returns to the surface, the copious fresh air pumped through continuously has removed every trace of the fumes. The paraffin is completely consumed, and the lamp itself is at the bottom of the sea.

"Mariners are right to fear fire. Oil, paraffin, fat: each kilogram burned releases a fearsome forty million Joules or so of heat as it is consumed. A few liters are enough to raise the temperature of a massive iron vessel above that which life can support."

"It is most ingenious," I had to admit, despite my annoyance.

Professor Challenger nodded. "I have no doubt, Doctor, that by this time tomorrow my explanations shall have proven correct, and superstitious talk will be stilled."

THE MORROW ARRIVED sooner than expected. I was woken from a deep sleep by frantic shouts above. Hastily putting on my outer clothing and grabbing a life-preserver, I made my way on deck. An eerie scene greeted me. Our propeller was turning at maximum revolutions, as evident from the thrum of the engines. But the ship stood almost stationary, in calm water just inside the mouth of a fjord.

There was near panic among the crew, and I felt a certain smugness: I alone of those present thought I knew the cause of the phenomenon that afflicted us. But a moment later, Challenger himself appeared on deck. He was carrying a lighted candle and a narrow-necked glass jar whose base appeared to contain lead shot. Ignoring the commotion about him, he knelt on the deck and inserted the candle in the jar, then stoppered it firmly. He leaned forward and dropped the arrangement gently overboard.

I rushed to the rail. The gleam of the candle was clearly visible through the water as the device sank. It was a most ingenious demonstration. If Challenger's theory was correct, the light should be carried sternward at great speed as it entered the hidden current.

For a moment, this seemed to be happening. The gleam moved sharply sternward as it sank. But a second later it seemed to pause and become fainter. Then it moved forward some meters, became brighter, and seemed to return almost to beneath me before I lost sight of it beneath the hull.

I looked at Challenger. He appeared even more baffled than I was: for several seconds his mouth literally hung open. Then a thoughtful look came into his eye, and without a word he ducked below.

I followed him down to his cabin. This was in a most untidy state. No attempt had been made to clear up yesterday's experiment: the contents of the glass tank slopped around as the ship rolled, the water undulating beneath the paraffin.

"You have been sitting here amid this mess?" I asked. He raised his eyes to meet mine.

"Yes, and thinking with my eyes resting on this tank the whole time. Talk of ignorance in the face of the most blatant hint. I have been blind indeed! And yet, what an extraordinary phenomenon to have discovered." He shook his head.

"Tell me, what do you observe in this tank?" he said.

"I see a layer of pink paraffin, the surface of which has ripples, as a result of the ship's motion."

"And what about the *other* surface of the paraffin?"

I was baffled for a moment; then I looked at the side of the tank, where I could see the boundary between the paraffin and the water below it.

"Why, there are much larger ripples on the surface of the water, upon which the paraffin rests. The interface is oscillating considerably."

"And can you deduce why that is so?" Challenger asked.

I shook my head.

"It is because to set any wave in motion, you must first give it some energy."

"Why, obviously: energy of motion."

"Not only that. There is also gravitational potential energy. Some of the fluid must be taken from the surface to make the

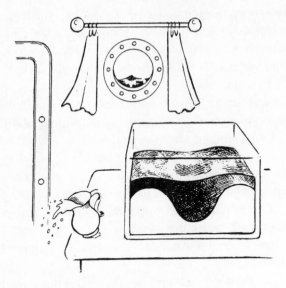

Professor Challenger's Cabin

troughs, and lifted to form the peaks. Now do you see? Since the paraffin is about nine-tenths the density of water, and rests on its surface, flowing down to fill the troughs and away from the peaks, to set up a wave of a given height at the interface requires many times less energy than if the paraffin were absent."

Call me stupid if you like—I am never at my brightest when just aroused from a sound sleep—but it seemed to me that his discovery was of little relevance to our voyage. I took myself back to bed.

I was awoken some hours later by a change in the ship's motion, and arrived on deck to find us tied up alongside the *Matilda Briggs*. I could see furious activity aboard her, seemingly at both the fore and aft ends, while our own vessel appeared deserted. I made my way to the gangway rigged between the two vessels, crossed with some trepidation, and asked the crewman guarding the far side what was afoot.

"It is that Professor you have brought with you, sir," was the disrespectful reply. "He is giving orders to lower the diving bell forthwith, and as for himself, why, he is about to leave on a bicycling trip, a quarter of a mile from land."

I made my way aft. The diving bell, a rusty shell about the size and shape of a cathedral church bell, rested on a raised trestle deck. At that moment, Professor Challenger scrambled from underneath.

"Ah, Doctor, you join us in good time. The bell is just about to dive; I have given the Captain my assurance that the mishap cannot recur, and checked with my own eyes that there are no sources of flammable energy, nor indeed anything else save two volunteers, aboard. All it requires now is for a witness of unimpeachable reliability to observe that nothing further is taken aboard now, nor anything untoward permitted to happen during the dive. I am sure I can rely on you. Allow me to lend you my spyglass."

I found a convenient spot to keep watch from, and turned the glass down upon the water. To my surprise, it was extraordinarily clear: I could see small fish scurry a good three meters below the surface.

"The water is most unusually transparent, for sea-water."

"Indeed, Doctor; I feel sure you will be able to deduce why."

Challenger picked up a small polished bucket attached to a rope, tossed it overboard, and pulled it back half full of water.

"Pray taste it."

I did so gingerly, and was astonished.

"Why, this is fresh water!"

"The glacier feeding the fjord starts to melt at this time of year. Still, it would be surprising if fresh water sufficient to fill the entire fjord was generated, would it not?"

"Surely it must have been, since we are moored quite near the center?"

Challenger smiled sunnily. "If you say so, Doctor. Now I have a mission ashore. I charge you to keep your post until I return."

A few minutes later I heard the splash of oars, and a rowing boat came into view manned by three crewmen. In the stern sat Challenger, with the dignity of a monarch on his throne. In the bow lay a bicycle, of the modern type with inflatable tubes for tires.

My attention returned to the foredeck as a party of crewmen energetically turned the winch handle, and the bell rose a foot clear of the deck. The hoist turned, and the bell was allowed to slide down to touch the calm surface of the water. The word *Sumatra* was painted round the top; I asked the hoist operator what it signified.

"Merely that she was built there in Indonesia, where she saw service for some years. She looks rather like a rat hanging by its tail, does she not? The Giant Rat of Sumatra, we call her, begging your pardon, sir."

The bell was now lowered very gradually, as she would be from here to the sea bed. All was quiet aboard, save the giant pump whose piston rose and fell massively, pushing a copious supply of air down to the divers.

I still felt indefinably uneasy. I could see the bell clearly beneath the water: nothing could approach unseen. How could the terrible accident possibly recur?

Was phlogiston really ruled out? Admittedly, it could not be a conserved fundamental quantity. But did that mean it could not exist? After all, Challenger himself had conceded that the waves I saw on the sea surface had real existence in some sense, and water waves were constantly coming into being and disappearing again everywhere. But then again, if clouds of free phlogiston were prone to form and wander around, such a phenomenon would surely have been observed before now.

At root it was a question of energy, that was clear enough. What kinds of energy were there? Thermal energy, yes, but the sea was cold, and heat never flows spontaneously from cold to hot. Chemical energy? I had the Professor's word there were no combustible chemicals aboard the bell. Elastic energy? A clock spring could hardly contain enough to do the job.

Mechanical energy? There were no mechanisms aboard the bell, but the ship herself had powerful engines. At the moment her propellers lay still, her boilers cold. Even if other sources were aboard, how could their energy be transmitted down to the bell? After all, the only connections between the two were a metal chain and a hollow rubber tube.

Somewhat relieved, I turned my attention away. The boat had reached the shore, and Challenger and bicycle were disembarked. I used my spyglass to watch as Challenger mounted, wobbled a few yards, then dismounted to connect a small pump to the rear tire, which he drove energetically to and fro.

After once more checking the bell, I looked shoreward again to see Challenger waving and shouting furiously at the boat. The crewman beside me looked on with amusement.

"Never a scientific mind so great but that he absent-mindedly left something behind, eh, sir?"

The boat turned back, and grounded near Challenger, who ran up and spoke urgently to the coxswain. To my astonishment, the latter fairly leapt ashore, leaving his mates adrift, and commenced to wave his arms in the manner of a demented scarecrow.

"It is semaphore, sir, and he is signaling distress," said the hoist operator. I passed the spyglass to him. He watched for a few seconds, turned pale, and barked orders at the winchmen.

"We are to haul the bell aboard at once, or death will result," he cried. "But how can he know that?"

Some ten minutes later, a breathless Challenger stood by my side as the bell was winched aboard. The divers scrambled out, baffled but unhurt. Challenger turned to me.

"There are times, Doctor, when even the greatest mind must be grateful to Providence for some timely hint. You saw me wield a pump ashore just now. Tell me, Doctor, do you own a bicycle?"

"I do not, but I have ridden them, and inflated the tires on occasion."

"Tell me, when you pump a tire up vigorously, have you ever noticed something curious about the metal valve connection?"

"Why, yes, it becomes hot, although I cannot imagine why."

"Why? Because a force in motion does work!" He pointed to the great pump on the stern. "Work equals force times distance. As the piston moves down, it does work upon the air. Air is compressible, and work is done as it is squeezed, as surely as when you wind up a spring. And where does the energy go? Where all waste energy goes: into heat!

"As the bell gets deeper, the force required to squeeze the air becomes greater, and so does the corresponding heating. The air streaming into the bell becomes tepid, then warm, then hot. With old-fashioned hand-operated pumps, the heat was not sufficient to warm the bell itself, but a steam pump capable of many horsepower—"

"Brilliant, to have seen the danger in time!" I exclaimed.

"Cretinous, to have overlooked it!" retorted Challenger. "But better in the nick of time than not at all. Here is a lesson which future engineers must learn, Doctor: that a profound and imaginative understanding of the laws of physics is as essential to the practitioner of the new craft as muscles were to his predecessor the blacksmith."

CHALLENGER DID NOT rest upon his laurels. He was rowed ashore again, and I watched as he embarked along the coast road. As dusk fell, he returned, accompanied by a cart full of some objects which turned out to be glass carboys, similar to but larger than the jar I had seen him fling overboard before.

"In an hour or so, the steamship *Scipio* is expected to bring us Professor Summerlee, who will state that all strange phenomena reported are attributable to ignorance or downright dishonesty," he remarked. "I think it is our duty to greet such a distinguished visitor. Will you accompany me, Doctor?"

And so I took the oars as we embarked together in the ship's boat. I had difficulty concentrating on my stroke as I watched the Professor's antics. He had by him a pail of water dipped from beside the boat, a stack of carboys, and a small charcoal stove glowing on the thwart. Repeatedly, he took a carboy, used tongs to insert a glowing piece of charcoal from the brazier, and placed the carboy in the pail. He sprinkled in lead shot until the neck of the carboy was floating level with the water surface in the pail. Then he removed the carboy, corked it with a glass stopper, and dropped it over the side of the boat.

I was highly amused to see that his great mind had overlooked one little detail. Presumably, he wished the carboys to float; but he had neglected the extra weight of the stoppers! And so each carboy slowly sank, unnoticed by him. I was tempted to point this out, but feeling that a little humility might do him good, I held my tongue for a while. As he prepared to deploy the last carboy, I gently pointed out his error.

He snorted. "My dear sir, you must credit me with the wits of an orangutan. Look back along our course, and tell me what you see."

To my surprise, I could see a row of orange dots. "Why, each seems to have come to rest at the same depth, just ten feet or so below the surface."

"Ten feet? Three meters, if you please! Now why is that?"

"Ah, I know. It is due to the compressibility of water, making it denser with depth."

"Hardly that. The compressibility of water is so low that it is barely a thousandth of a percent denser at that depth than at the surface. Quite another effect is responsible. Recall that sea-water is about two percent denser than fresh. If you poured fresh water onto sea-water so gently that the two did not mix, what would occur?"

"It would no doubt be like the paraffin. A layer of fresh water on salt would result, although the boundary would be invisible."

"Invisible unless you drop a row of beacons whose weight is judged so that their density is just greater than fresh water, just less then salt, to mark it. But look! Here comes the *Scipio*, ahead of time."

I followed his pointing finger, and saw the *Scipio* slide into the fjord at a good speed. But shortly she seemed to falter, and at the same time I beheld a most extraordinary thing. Under water, the red dots of glowing charcoal were undulating in synchronized motion, for all the world like a giant sea serpent writhing in the depths.

"Good Lord," I cried, "any fisherman who believes in *Krakens* will surely be fleeing for dear life."

Challenger smiled condescendingly. "Have no fear, my dear sir. Tell me, when a ship starts her propeller churning, where does the energy go?"

"Well, initially, into accelerating the vessel."

"And when she reaches full speed?"

"Into pushing the water aside, no doubt."

"Yes—or, in effect, into creating waves. The wake of a ship is but artificial waves raised by the action of her engines. Do you remember how the tank of two liquids, paraffin and water, allowed

The Hidden Wave

much larger waves to be created at the interface than at the surface? In fact, the nearer the density of the two liquids, the greater the height of the waves at their boundary. And with a difference of just two percent—"

"The invisible wake at the boundary will be vastly greater than that at the surface," I exclaimed.

"So much more so, that it steals all the vessel's energy, and the ship almost comes to a stop, for all that the engines can do," said Challenger complacently. "No doubt the learned Professor Summerlee should be able to deduce this detail. But just in case, perhaps we should row across and greet him, before any superstitious crewmen mutiny and fling him overboard."

He sat back and watched benignly as I laid my weight into the oars. "You rowed at school? Commendable to see such robust energy at work. And a little symbolic, Doctor, for we have demonstrated that energy itself is a robust concept. Whereas phlogiston and epicycles evaporate like mirages before the clear-sighted investigator, we have seen that even when energy *appears* to be created or destroyed, it is not so. Have the faith to look for subtly concealed forces, such as invisible waves, and lo! we find them. Energy is a concept worth hanging on to, even in the face of the odd little local difficulty."

3

The Case of the
Pre-Atomic Doctor

"A HARD DAY'S ROUNDS, DOCTOR?" inquired my friend solicitously as
I laid my medical bag wearily on the side table with a longing glance
toward the decanter.

This kind of sympathy is always welcome from a fellow toiler,
but grates somewhat when it comes from a man who, still adorned
in dressing gown and slippers at five-thirty in the evening, has spent
his day playing idly with that grown man's toy, one of his frequent
chemical experiments. However, I had indeed suffered that afternoon, and the temptation to utter some sharp retort vied and lost
with the need to unburden myself, perhaps in the hope of some
sympathy and advice.

"Well, in fact, Holmes, a routine enough day until my final call:
a middle-aged lady with stomach pains of a longstanding nature.
And that not a baffling case: if she would listen to my opinion and
consent to a minor operation, it would be very much in her interests.
But her mind is closed to my advice."

"She distrusts the medical profession?"

"No, she believes we doctors have our place. The problem is

that she is under the influence of a most sinister quack, a man she has been introduced to who believes in crystal healing and homeo-pathic remedies. I saw his carriage, a fine two-horse brougham, pull away as I arrived for my visit. He is a wealthy man, well dressed, with an imposing look to him—much more impressive than myself, I could not help but feel."

"No doubt his fine clothes and carriage were financed by many a gullible patient! So what does this charlatan prescribe for her?"

"Well, he claims that her problems are caused by toxic sub-stances she must have eaten as a child. And his remedy is quite alarming: he claims that taking a greatly diluted solution of these poisons—lead, arsenic, belladonna, every common poison you can think of—will cure her. And unlikely as it sounds, such is her belief in him that she has started taking this prescription daily."

"It certainly sounds likely to worsen things rather rapidly. This could almost become a case for me, Watson! When you say his medi-cine is well diluted, to what degree of watering down do you refer?"

"Quite a lot, or I would be more worried than I am. He claims that after starting with a beaker of the pure poisons, he pours nine-tenths away and replaces its volume with water. He then repeats the procedure, to turn the one-tenth solution to one-hundredth, and so on for thirty separate stages."

"That is certainly taking caution to extremes. A tenfold dilution, repeated thirty times over. So the ratio of water to poison in the final beaker would be written as one, followed by thirty naughts, to one. A one-million-million-million-million-million-fold dilution. I think the scientists' way of expressing such a figure, ten to the thirtieth power, is really more convenient. That is a dilution equivalent—let me see, if his original beaker contains one liter of water, he would have to mix it with a volume one million kilometers on a side—why, Watson, his solution must be far more dilute than if he were to pour his beaker into the Pacific Ocean, stir well, and fill the beaker again from the resulting mix. At least you can be assured the potion is harmless!"

"Well, that is all to the good, but nevertheless unless I persuade the lady to accept more effective treatment, all may soon be lost."

Sherlock Holmes frowned, and sat silent for a minute or more, his fingers steepled. Then suddenly he flung his head back and laughed.

"Upon my word, it looks as if my time today has not been so idly wasted as you suppose. No, don't bother to deny it: your look at the sight of my day's activities was plain enough. But do you know, Watson, what I have actually been doing?" He picked up a scientific monograph I had noticed beside his chair.

"I can deduce that it involves an eye-dropper, some dishes of oily liquids, and considerably less mess and stink than your experiments normally produce," I said. "Have you taken up the manufacture of beauty products, perhaps?"

"No, Watson, that is well wide of the mark even for you. I have been repeating some recent experiments which verify, pretty much conclusively, the existence of atoms."

"I had thought it was long accepted that matter must be made up of tiny indivisible particles, albeit far too small for the most powerful microscope ever to resolve?"

"By no means, Watson; even now quite a few scientists continue to doubt. The evidence for the so-called atomic theory has been very much circumstantial. One strong clue is the existence of crystals. If almost any substance is placed in a liquid solution, and then gradually forced to resolidify—for example, if it is dissolved in water which is then slowly evaporated—it tends to condense in the form of regularly shaped crystals. By and large, any given chemical shows a strong preference for one particular shape of crystal. It suggests that the substance is made up of many microscopic units of just that same shape, individually too small to see.

"Another clue is the tendency of chemicals to react together in combinations of exact numerical ratio. For example, to make water, burn one weight of hydrogen with eight of oxygen. To make methane, burn one weight of hydrogen with three of carbon. To make carbon dioxide, burn three weights of carbon with eight of oxygen. And so forth. The ratios hold very precisely, and the obvious explanation is that tiny units of these indivisible elements—no one has ever succeeded in breaking hydrogen, carbon, or oxygen into constituent parts—combine in fixed ratios to form larger compounds."

"I have always found such reasoning convincing," I said a little smugly, for I remembered the lesson well from my basic medical training.

"But this evidence tells us nothing about the actual size of atoms, beyond that they are tiny," said Holmes. "Watch now. I take a dish of pure distilled water—"

"Almost as pure as that quack's medicines," I jested.

"—and an eyedropper full of oil. I squeeze a tiny drop of oil onto a wire, and hold it before this ruler. Take that lens, Watson—no, the more powerful one—and tell me the dimensions of the drop."

"It is just about one-fifth of a millimeter across."

"Good. Now, I place the drop upon the water, and—"

"It has vanished, dissolved, I should say."

"No, Watson, oil does not dissolve in water. The substances are immiscible, and the surface tension between them draws the oil into the thinnest possible film upon the surface. Squint at a shallow angle."

"Ah, I can just see the film. Only as a faint shimmering, but I can judge the edges. It is a spot about ten centimeters across."

"Well observed, Watson. Now, you can work out the dimension of an oil molecule for yourself. You have only to divide the volume of the oil drop by the area covered, to work out the thickness of the layer."

I did so, with a little help from Holmes, who reminded me that a rough estimate of the volume of a sphere is half that of the cube which could contain it, and of a circle, three-quarters of the square which could contain it. But I thought I must be mistaken.

"It comes out at a half of a thousandth of a millionth of a meter, Holmes. Surely that cannot be right: such a tiny result, from quantities big enough to see with the eye."

"It is correct, Watson. But more conveniently expressible. If one thousand million, written as one followed by nine naughts, can be spoken as ten to the ninth power, then one thousand millionth is just the inverse: ten to the *minus* ninth. So in scientific parlance, the width of the layer is roughly point five times ten to the minus nine, or equivalently five times ten to the minus ten meters.

"And there you have a figure—a crude and disputable figure, I admit, but at least an estimate—for the size of one molecule of oil. Thirty million side by side would be required to span the width of your thumb."

I sat back in my chair with a sigh. "You have temporarily diverted me from the cares of my day, Holmes, but have you not forgotten my problem?"

"Not a bit, Watson. The point of the exercise is this: How many molecules would fit in a test tube containing, say, one hundred cubic centimetres?"

I calculated. "Ten to the eight cubed. Ten times eight to the power of three—why, that is ten to the twenty-fourth. One million million million million. Although why that is important is beyond me."

"Now what factor of dilution is your competitor using?"

"Ten to the thirtieth. Why—just a moment, Holmes—that is a number greater by far than the number of molecules of poison he must have started with."

"So much so, Watson, that there is but one chance in ten to the sixth—one chance in one million—that even one molecule of the original poison remains, for good or ill. You have but to explain this reasoning to your patient, and the light should dawn that this man's nostrums are nothing but water washed down with smooth patter. It is a paradox he will find hard to resolve, that a medicine containing not a single molecule of active substance can somehow effect a cure!"

IF HOLMES HAD more opportunity to meet hypochondriacal and otherwise difficult patients, he might become less sure of the power of logic to persuade. I returned to Baker Street the next day with a heavy heart.

"My efforts were to no avail, Holmes," I said as I took off my coat. "Doctor von Kranksch—for so he styles himself—was already at her bedside when I called. She insisted on him staying through my visit, and he made mincemeat of my arguments. Why is it, Holmes, that clever words seem to carry so much more power than logic?"

"If I could answer that one, Watson, half the world's problems would be solved at a stroke. But come sit down by the fireside, my dear fellow, and tell me in detail what went wrong."

I took my chair, and stretched my legs wearily.

"Well, he disputed the existence of atoms. First he discounted the oil-drop argument. He claimed the thickness of the surface film might be merely a consequence of the way the attractive forces between the two liquids worked. At all events, the most the experiment proved was that *if* atoms existed, they must be smaller than the thickness of the film.

"I tried to argue, Holmes, but somehow my words carried insufficient conviction. Eventually I had to leave. To convince my patient, I would have needed some quite direct and dramatic demonstration of these atoms, and especially some way to measure their actual size."

Holmes sat forward eagerly in his chair. "Then I am your man, Watson. What would you say if I told you that you could peer down that microscope and see evidence of atoms in plain sight?"

He waved a hand toward his battered microscope. I saw that it had been modified so that in place of the objective slide, a small tube, apparently of empty air, had been fastened. I went over and bent my eye to the lens. With astonishment, I saw distinct black specks dancing in the field of view. They were almost large enough for me to discern details of their shape.

"This is wizardry, Holmes! I thought I had read somewhere that atoms must be a thousand times too small to perceive even through the most powerful microscope that could ever be made."

"And so they are, Watson."

I raised my head and looked suspiciously about; it is a shame when Holmes uses his admittedly greater intelligence to pull my leg. I noticed a flask of grayish dust by the microscope and bent to read the label.

"Really, Holmes! What is the point of showing me grains of pollen and telling me they are atoms? It is no joking matter when my patient lies ill for want of better advice."

"I am not making fun of you, Watson. Yes, those are grains of pollen that you see. But what do you observe about them?"

I bent my head to the microscope once more. "It is hard to discern any detail, Holmes. The things dart about so, and will not stay still."

"Exactly, Watson! Now, why do they jump about like that?"

"Is it the effect of the light? Or do pollen grains perhaps have flagellae, like bacteria, which they wave to move themselves about?"

"No, these possibilities are easy to disprove. It is something more fundamental. You know that heat is in reality motion: it is the vibration of the substance of a solid, or the continuous movement and colliding and rebounding of the substance of a gas. That is why air exerts pressure, and why it will rush spontaneously to fill a vacuum.

"Now, Watson, suppose for a moment that air molecules, although invisible, were in fact very massive, with correspondingly few of them in a given volume. Might the effects of their motion be observed?"

"In principle, yes, as a tingling on the skin. In the reductio ad absurdum limit, I suppose the vibration would become enough to bounce you about as you walked or sat! But of course they are really so tiny, Holmes, that the average of huge numbers of molecules hitting you each second is to a good approximation constant, so we perceive the air as having a steady pressure."

"Very good, Watson. Now consider the pollen grains. If an air molecule were as massive as a pollen grain, then bearing in mind that air molecules bounce about at some one thousand miles an hour, what would you see?"

"Why, only a blur. You would never be able to spot an individual grain in such violent motion as the repeated collisions would cause."

"And if air molecules were truly infinitesimal—billions of billions of billions of times lighter than a pollen grain?"

"Then the impacts would average out almost perfectly, and the grains be still."

"Exactly, Watson. And by measuring the motion, which is in fact intermediate between these two cases, it is possible to deduce what the ratio of the mass of an air molecule to a pollen grain must be. By weighing a large, counted number of pollen grains, you can find the mass of one grain, hence the weight of the average air molecule. And knowing from the observations of chemical combustion that air is eighty percent nitrogen, each nitrogen molecule being composed of two atoms—"

"You have a precise measure of the weight of an atom! Holmes, you are a genius."

Holmes smiled. "This is not my original experiment: I am merely following a description in the journal you see there, of work done in Germany. It turns out that atoms are about two times ten to the minus ten meters in diameter, or one five-millionth of a millimeter, if you must use unscientific language. But the point is, Watson, that you can tell your lady client truthfully that the size of the atom is known, and that direct evidence for the existence of atoms can be seen."

THE FOLLOWING EVENING, Holmes seemed deep in thought when I entered. But as I drew up an armchair, he glanced at me alertly.

"Well, Watson, how went it?"

"It has been touch and go, Holmes! I visited the lady twice. This morning, I argued with her at length, until eventually she instructed me to leave. Calling back this evening, I feared the worst.

"But she welcomed me in, and told me she had been pondering my arguments all day. She told me that she sat by the window, and through the edge of her reading glass, she could all the time see grains of dust dancing in the air, illuminated by the bright sunlight, reminding her of my words.

"She said that for a while she sat baffled by the paradox: she believed that her medicine was proven to work, yet my reasoning had shown her it could contain not one atom of the supposedly useful poisons, so it could not possibly be effective: a baffling contradiction.

"Then, she told me, she remembered previous times when an apparent paradox had puzzled her. She said each time in her life she had come upon apparently contradictory facts, it was almost always the case that some new, proven observation clashed with some assumption—sometimes an implicit, almost unconscious belief—which although deeply held, had not in fact ever been proven beyond doubt. It was really the mental pain involved in reordering her thoughts, and grasping that she had been mistaken, which tended to prevent her from abandoning the old and accepting the new.

"No doubt you think her a most foolish woman, Holmes, with little ability for clear thought."

"On the contrary, Watson, I wish that half our so-called savants had her ability to face new facts."

"At any rate, she realized that although Doctor von Kranksch had told her much about the almost miraculous success of his medicines, and she had trusted him as the layman tends to trust those who claim expert knowledge, she had no proof of his claims. At last she told me that she had finished with von Kranksch and would take my advice. She has finally seen the quack for what he is, and even insisted I take away the medicines and pamphlets of his she still had, to reassure me she had no further use for them."

I pulled the material from my bag as I spoke. I had intended to cast it into the fireplace, but Holmes held out his hand for it.

"This is interesting, Watson," he said at length. "There is no law, alas, against merely telling lies. But claiming false medical qualifications—there are some curious things here—well, it is hardly my worry, Watson, but next time Lestrade calls, you might just remind me to give him these." He scribbled some notes on the pamphlet, and slid it behind the letter rack.

LATER THAT EVENING, I took a deep breath.

"There is just one thing still troubling me, Holmes, although you will think it absurd. Yet you seemed to agree with my patient's philosophy that one should always be willing to question comfortable assumptions in the light of new evidence."

Holmes nodded encouragingly.

"Where I really came unstuck arguing with von Kranksch was on the subject of crystals. I wished I had never brought the topic up! He claimed the atomic theory of crystal shape was unproven. He said that many substances will not form crystals at all, and some are capable of forming crystals of two or even more differing shapes."

"That is quite true."

"But then he went on to claim that crystals had mystical properties that today's science would never understand. He says that every crystallizable substance resonates with itself in a mysterious way not constrained by the normal boundaries of space and time. He claimed that the proof is that when any new chemical is first isolated by scientists, it is very hard to make it form a crystal. But when a second experiment is performed on the new substance, crystals form far more easily."

"That is not hard to explain, Watson. It is well known that the presence of a seed crystal or crystals in a solution—even microscopic ones too small to detect except by inference—greatly facilitates the development of others. So crystallize a substance once: there will soon be microscopic traces of it everywhere about your laboratory. When you repeat the experiment, even though you think you have destroyed every part of the earlier sample, behold— the second attempt is miraculously easier!"

"I can see how that would work, Holmes. But he claims that even if the second sample is made quite independently, say, half a world away—the first in England, the second in Australia—the effect still holds. He claims a mystical field permeates all matter, perhaps modified by the minds of observers such as the scientists present, and so when a crystal of a given kind has formed even once somewhere in the world, it becomes easier for another like it to form on each subsequent occasion, imitating the first. Surely he must be lying about this so-called resonance effect?"

To my astonishment, Sherlock Holmes shook his head. "There is documented evidence of such a thing, Watson. But do not trouble yourself too much. First, when a man does an experiment following the instructions of another, who he knows has succeeded before him, the experiment tends to be done in a swifter and more confident way. And anyone is quicker to notice something when he already knows to expect it than when he searches for an unknown. So there are psychological factors which could account for the phenomenon.

"But there is a more intriguing explanation, which has been

described in the literature rather charmingly as 'The Puzzle of Caesar's Last Breath.' It brings home rather vividly the true implications of the tiny size of atoms.

"Imagine you were standing in that Roman chamber as Caesar gasped: 'Et tu, Brute.' In your opinion as a medical man, Watson, what would the actual quantity of air in that last gasp of his have been?"

"At least a liter, Holmes. The volume of an inhalation can vary greatly; it depends on a number of factors. But really, what has this to do with crystals or atoms?"

"Bear with me, Watson. The density of air is about one point two kilograms per cubic meter, so it would be fair to say that his last breath had a mass of at least a gram. Now, how many grams of air make up the entire atmosphere of this planet of ours?"

"Really, Holmes, that is the kind of specialized information which a layman such as myself cannot possibly know, or even find out at all easily."

"Ah, but you already know it, Watson! What is the diameter of the Earth?"

"Eight thousand miles, almost exactly."

"And the pressure of air at the surface?"

"Fifteen pounds per square inch."

"And there you are! You have but to multiply the area of the Earth's surface, in square inches, by fifteen, and you have the total mass of the atmosphere in pounds. But you will find it easier using the Continental metric system of measures. I will give you a start: the Earth's surface area is about five hundred million square kilometers, and the pressure of the air works out at a convenient one kilogram per square centimeter."

"Well, Holmes, there are one hundred centimeters in a meter, and one thousand meters in a kilometer—"

"Scientific numbers, if you please. It will be easier, take my word for it."

"Then there are ten to the two centimeters in a meter, and ten to the three meters in a kilometer, so ten to the five centimeters in a kilometer. Why, Holmes, to multiply these numbers, all you have to do is add the powers of ten—add the number of naughts, in effect."

"A remarkable discovery, Watson! Pray continue."

"So there are ten to the five times ten to the five—that is, ten to the ten—kilograms of air on each square kilometer. Multiply by five times ten to the eight, and you have five times ten to the eighteen."

"Yes, but that is kilograms, Watson—never forget the units you are using."

"Times ten to the three—so five times ten to the twenty-one grams, or last breaths. In millions—"

"No, Watson, do not translate. The essence of mastering a new language is to continue to think in it. Now, I will tell you a remarkable fact. Air molecules are so small that each weighs just five times ten to the minus twenty-six kilograms. So how many molecules were there in that breath?"

The division took me a moment longer. "Two times ten to the twenty-two, or twenty times ten to the twenty-one. Why, that is actually four times the number of breaths making up the whole atmosphere."

"And if you think about it for a moment, Watson, you will realize that with every breath you take, on the average you inhale four molecules from Caesar's dying gasp!"

"Now you have succeeded in making me feel quite queasy, and I ask you again what on earth this has to do with crystals?"

"Well, Watson, suppose now that I make up a test tube containing say ten grams of some strange new substance. Suppose further that I absent-mindedly leave it upon the window sill until part has evaporated."

"There is no supposing about it, Holmes, and if you do not remedy your habits, one of those days Mrs. Hudson, tolerant of you as she is, will undoubtably—"

"Now, Watson, allow a few days for atmospheric mixing to take place. How many molecules will there be in each liter of the world's whole atmosphere?"

"Why—good gracious, Holmes, your vile mix will have polluted every breath of air upon the planet."

"And if a chemist in Adelaide should try to make the same chemical crystallize?"

"This is incredible, Holmes, but it would seem molecules of your experiment will be falling from the air into his retort, and quite feasibly seeding the formation of his crystals!"

"Indeed, Watson. And the odds are still further improved if in addition to the random mixing of the atmosphere, there has been some more direct contact between the laboratories—a parcel sent from mine to his, for example, whose surface will inevitably be contaminated with many millions of atoms. There is always traffic between the world's centers of learning."

I pondered for a while. It occurred to me that I owed my friend an apology.

"I must confess, Holmes, I had thought there were few more futile questions than speculations about the existence of atoms. Given that they are too small ever to see, I had classed atomic discussions with arguments about the existence of life on Mars, or whether the chicken or the egg came first: unresolvable conundrums whose solution was in any case of no importance. I thought your puttering the last couple of days a complete waste of time for a grown man with work to do."

"In a sense it was, Watson: the work had already been done more skillfully, and my talents are for detection rather than the pure sciences." Holmes smiled. "But as a medical man, you should have guessed the matter might be important. Have you not heard of the miasma theory of disease?"

"Why, of course, Holmes. At the hospital where I trained, many if not most of the older consultants believed it. It has been known since ancient times that disease could pass from one person to another; thus there must be a transmissible agent involved. It was suspected that this consisted of some kind of intangible field or gas, called the miasma."

"A little like phlogiston, eh, Watson?"

"I can see the analogy. But of course another theory was that disease is caused by tiny parasitic organisms; and that is now certain, for they can be clearly seen in modern microscopes, and we call them bacteria."

"So the question of whether disease was in essence an indivisible substance or miasma, or made up of tiny similar discrete bodies, was of some practical importance?"

"Of enormous importance, Holmes: all the hopes of modern medicine depend upon it. But I see you are pulling my leg, and that is unfair of you, for have I not just admitted that atoms are important also?"

"You have conceded it handsomely." Holmes rose from his chair. "But there is a further moral to remember. The arcane question of atoms turned out to be important not merely to philosophers of science but to a quite ordinary woman with no previous interest in such matters. Had she not understood clearly, it might well have cost her her life. Good night, Watson."

4

The Case of the
Sabotaged Scientist

"COME, WATSON, AND TELL ME what you make of this gentleman."
I hurried over to stand beside Holmes at the window.

"I thought you had no appointments for today," I commented.

"Indeed, but look at the man on the opposite pavement. He certainly has the look of a client, and an excited one at that."

Holmes pointed to a man standing by a hansom cab, from which he had evidently just dismounted. He was engaged in some sort of altercation with the driver. A moment later, the cab pulled away and I could see him clearly: He made a somewhat comic figure—a tall, lean, scraggily built man with a long growth of untidy beard. He wore an expensive-looking dark suit, but so creased and rumpled that it hung oddly even from this distance. The man commenced to frantically rummage through his pockets, turning up scraps of paper each of which he inspected through a gold pince-nez, then tossed aside.

"Normally, I pride myself that I can tell most men's walk of life at a glance, Watson, but the indications here are certainly confusing. What would you say?"

I tried to emulate my colleague's methods. "He is clearly a man of means, Holmes: not only is the suit expensive, but the shoes also. But to have left the house so disheveled, he has clearly suffered some great and sudden shock—no doubt the matter which brings him to see you."

"Not bad, Watson; you did well to notice the shoes. A smart suit with cheap uncomfortable shoes generally reveals a man posing above his station. But it is not a crisis of this morning alone which has caused him to leave an expensive suit crumpled at the bottom of a closet for weeks, or to delay a haircut for a similarly unwise length of time."

"Well, Holmes, he looks just like the popular caricature of a mad scientist. I could visualize him in one of Mr. Wells's romantic fantasies. So of course he cannot be that. Frankly, he acts more like a refugee from an insane asylum!"

For at that moment the figure, having at last found the piece of paper he evidently sought and peered around at the street numbers, had darted out into the road with such suddenness he was almost mowed down by a cart-horse. He reached the pavement on our side more by luck than by judgment.

"No, Watson, I should say rather—ah, but I was wrong: he is not a client at all, and our competition is void."

For the man rang not at our door but the adjacent one. Sherlock Holmes turned away from the window.

"Well, Watson, that is something of a relief. I had arranged a day clear of appointments because I particularly wished—"

"You spoke too soon, Holmes."

For the man had emerged from the next doorway, gesturing impatiently at the occupant; then the peal of our bell was heard.

"Confound it, Watson. Perhaps Mrs. Hudson will deflect him."

But the hope was vain: we heard a shrill voice downstairs, and moments later the man was being ushered into our presence.

"Is one of you Mr. Holmes, the famous detective?"

"I am. But I normally see clients only by appointment, except in grave emergencies."

"Then you will certainly see me. I am the victim of a crime whose importance can scarcely be exaggerated."

"And you are Mr.—?"

"Doctor Illingworth, to you, sir, if you please! Doctor Illingworth of Edinburgh, at present seconded to Cambridge University."

I recognized the name of the man rumored to become the next Astronomer Royal, and evidently my companion did as well. His manner grew a trifle less impatient.

"If you will take a seat, Doctor Illingworth, and endeavor to explain the situation calmly, I will see—"

"No time, sir! Minutes are vital, or the evidence could be gone. The scene of the crime is the British Museum, just a few hundred yards away: I will explain as we walk."

I could see that Holmes still looked doubtful. As my eyes had recently been opened to the importance of matters scientific, I felt it my duty to intercede.

"I am sure Mr. Holmes will be able to help you, Doctor; he is a keen supporter of the sciences."

Shortly, we were striding briskly along the pavement, Sherlock Holmes still in none too good a temper at the interruption of his day.

"Now, Doctor, if you would be so good as to explain the events which have occurred," he said briskly. "What, precisely, is the nature of the crime?"

"In the mundane sense, sabotage. In the grander sense, an issue basic to the progress of science. A question of cosmic importance is at stake—a question which could affect our whole place in the Universe. More than this it would be unwise to say at present."

"I am beginning to think you were right in your first guess, Watson," Holmes remarked quietly to me. Then, in a louder tone: "Cosmic questions are slightly beyond my humble scope, Doctor. Let us take the mundane matter first: what exactly has been sabotaged?"

"My plates, sir. My finest and quite unique plates!"

"You have come to me because your maidservant has broken some crockery?"

"Your jest is in poor taste. I am referring to photographic plates.

I have found it necessary to count the number of very faint stars in the heavens, and for the purpose I needed a photographic material of exceptional quality. My colleague Doctor Adams, the director of the British Museum here and a leading chemist, was good enough to synthesize for me a photoactive substance of exquisite sensitivity. With it I coated some glass photographic plates of very large size, one meter square, and was ready to start my search."

He pointed upward. We had just turned the corner leading to the Museum, and following his indication, we saw atop it a feature I had never before noticed: a small cupola in which was mounted a telescope.

"You are conducting the search from the Museum's own observatory?" I asked.

"Do you take me for a fool, sir? The search proper requires a much larger telescope, away from city lights. But before dispatching the plates, I thought to take a test shot from the telescope here. And the plate was quite spoiled, I can only assume by some saboteur."

We were now on the Museum steps, but Holmes came to a halt. "You mean to say," he inquired in a dangerously quiet voice, "that you have called me out pleading urgency and importance, because a photograph did not come out properly?"

"Exactly! It is true what they say, Mr. Holmes, you are quick to grasp the essence."

My friend took a deep breath. "It is a pity, Doctor Illingworth, a great pity, but I have suddenly recalled a mission of even greater urgency. A lady in Brighton has requested my help in a poisoning case, and I must go there without delay. Sir, I bid you good day."

"But Holmes," I said, surprised at his lapse of memory, "she telegraphed last night to explain that the local police had already solved the matter. Surely I passed the message on to you."

Sherlock Holmes looked at me with disfavor, and I perceived that I might have blundered. But at that moment there came a hail from the top of the steps.

"Mr. Holmes, we are delighted to see you. How like your reputation to concern yourself with what might seem our trivial problems!"

Our greeter was the Director of the museum, Doctor Adams, a man both famous and known to us slightly. Holmes sighed, and allowed himself to be ushered inside. We were conducted to a hallway against one wall of which stood the plate described.

Expecting some mundane vandalism, I received a shock. The plate showed a clear and beautiful picture of the constellation of Orion the Hunter. But superimposed across the scene were a set of shadows of most sinister appearance. Vaguely anthropomorphic, although distorted and blurred, they reminded me of some cross between man and beast. I could almost believe this sensitive photography had revealed that there were, after all, pagan gods striding and battling in the depths of space.

"It is difficult to see this as chance damage," admitted Holmes. "But who could wish to do such curious sabotage, and for what sensible purpose. Who had access to the plates?"

"Foreigners, sir!" replied Illingworth immediately. "The observatory is kept securely locked, but because of its large size the plate was left down here for a time before being carried up the staircase to the cupola. The museum itself was locked for the night, but it would be easy for any visitor to have concealed himself somewhere on the premises at closing time."

"And the motive?"

"Scientific rivalry! Since the credit for the finding of Neptune was so disputed, there has been intense jealousy among Germany, France, and Britain. First discovery of things astronomical has become a matter of national pride. At all events, I have evidence that an intruder was present." Illingworth bustled away.

"Most likely some student prank, if sabotage at all," said Holmes quietly. "I can hardly imagine the good Doctor is popular with his pupils. What do you think, Director?"

Dr. Adams pursed his lips. "Well, it is difficult to see how this damage could have been accidental. But I agree a youthful prank is far more likely than some foreign plot."

At that moment, there came a nearby clatter of mops and pails. The Director sprang forward and hastily covered up the front of the plate.

"I do not want the cleaning staff to see this," he explained. "There has been much superstitious chatter since we received some strange relics last week, and this sight would undoubtably encourage it."

"Strange relics?"

"Yes, from the Dangerfield expedition. Have you not heard the story?"

We had indeed. The Dangerfield party had returned a few weeks before from a trip to Central Africa. They reported a ruined city, containing strange metal statues which appeared of sophisticated manufacture. It was suggested these must have been planted by some rival party as a hoax, but what was even more mysterious was the bad luck which had since befallen the explorers. Two had strange burns, most of the others had become sickly, and all now appeared to be wasting away although no known disease could be diagnosed. There had been much talk of curses like those said to afflict pyramid robbers, which Holmes had angrily pooh-poohed; I had suggested some new tropical disease was a far more likely explanation.

The Director indicated a table a few feet away. "We have placed one of the items on show. That idol is made of a rare metal almost twice as dense as lead, and is of interest even if it is not of ancient manufacture."

I walked across to examine the piece. It took the form of a hemisphere, with a face carved on the inner surface. The face was in inverse relief, with the features concave rather than convex, as for a mold. A trompe l'oeil effect made the face appear normal when seen from a distance, but the shadowing of the features changed unexpectedly as you moved about, so the blind eyes appeared to be following you. The overall effect was most sinister: I was not surprised that superstitious servants should be fearful of the thing.

At this point Illingworth returned, carrying a small box with glass sides. A copper strip protruded from the top and ran vertically down through the center. Attached to it was a flap of yellow metal, which swung freely in the interior.

"Gentlemen, this device is an electroscope. I charge it thus." He rubbed a piece of cloth briskly against his sleeve, then touched it to the top of the box. The yellow flap immediately sprang out at right angles to the central column.

"You are familiar with the concept of electric charge. Matter is made up of positive and negative substances which are normally perfectly intermixed. An electric current is a flow of the negative substance through the positive. A static charge is a slight excess of positive over negative, or vice versa. Rubbing this chamois leather expels some of the negative material from it. Like charges repel, and so the flap of gold leaf is forced away from the central strip, as both become equally positively charged, being electrically connected."

"I follow you, Doctor, but what has this to do with your sabotage?"

"Well, although it was purely accidental, the device made a rather neat burglar detector. I had left it charged on the table. In the dry air of the museum, left to itself the device will stay charged for many hours—unless I discharge it thus, by connecting it to the Earth through my body."

He touched a finger to the top of the box: the gold leaf immediately drooped limp.

"I thought nothing of it at the time, but I now recall clearly the device had discharged itself in the space of an hour or so while I was making adjustments in the observatory as the plate stood out here. Yet no one else should have been in the museum at that time. The clearest proof of an intruder!"

Holmes seemed quite cheered: he rubbed his hands together. "Gentlemen, I see the way to a solution of your problem. Watson, you are a firm respecter of the sciences, are you not?"

"Certainly, Holmes."

"And an admirer of this great museum who always relishes an opportunity to spend time amid the treasures here?"

"Indeed," I replied tactfully.

"Then it is settled. Illingworth, you could leave a fresh plate here as bait? Capital! And Watson will stand guard overnight with his trusty revolver. Be it foreign spy or student, he will certainly be able to detect and arrest the miscreant."

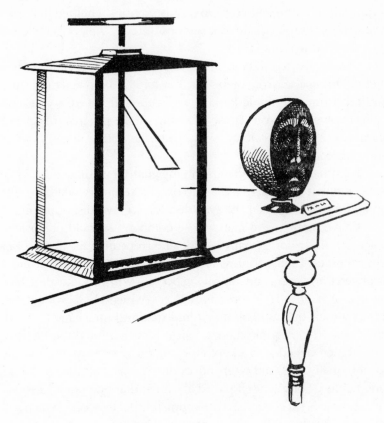

The Electroscope and the Idol

"But Holmes, my patients!" I protested hastily.

"He has been telling me how so many of his patients are now on holiday in these dog-days of summer that he is looking for ways to expend his time, and indeed that of his friends. He will be here at nine tonight. But for now we must bid you good day."

I FELT MOST self-conscious as I ascended the steps of the museum at dusk and looked in vain for a bellpush or knocker on the massive outer door. But someone was keeping watch inside, for I heard the noise of unbarring, and a small wicket gate opened. Illingworth greeted me perfunctorily and led me through the now darkened ground floor of the building. On his instructions, I helped him carry a new photographic plate to a site opposite the stairs which led to the Observatory cupola.

"There, Doctor Watson, all is precisely as it was last night. I had the staff bring a seat for you before they left." He showed me a small and most uncomfortable-looking wooden chair. "I trust you have your revolver with you. Those who imperil the progress of science should be shown no mercy, sir: have no hesitation in carrying out your mission."

I assured him I would perform my guard duty diligently, while reflecting that if he expected me to exact a summary death penalty on some practical joker, he was due for a grave disappointment.

I have rarely met a man whose company I would prefer less, but when Illingworth finally left me after repeating his instructions several times, I must confess to a certain chill creeping over me. I arranged the blanket and improving book—Winwood Reade's *Martyrdom of Man*, at Holmes's suggestion—which I had brought, but found myself quite unable to sit and read. All about me the dark skeletons of the terrible dinosaur lizards which ruled the Earth so long ago crouched, and I kept fancying I saw movement from the corner of my eye, although no doubt it was only the shadows lengthening. For a while I paced about, and at one point I rubbed my sleeve and succeeded in recharging the electroscope, which still sat on the nearby table: with memories of being told sternly never to touch museum exhibits, the exploit gave me a childish thrill.

Eventually I resolved to overcome my restlessness and settle myself for the night. I dragged the chair right in front of the giant photographic plate, its back to the wall so no one and nothing could approach unseen, and sat. But I immediately rose with a start: I had quite forgotten the idol head, which sat on the tabletop not two feet in front of me, and by its infernal illusion appeared to be staring right into my eyes. Almost, I was tempted to rise and cover it with a cloth, or even move it, but the thought of what would be said if I somehow damaged such a valuable item deterred me.

For a time, I distracted myself by staring up at the stars visible through the great high windows, and wondering idly what Doctor

Illingworth's new theory could be, but still I could not rid myself of the illusion that something in the room was moving or somehow changing, albeit with imperceptible slowness. Each time I lowered my eyes from the heavens, I told myself sternly the scene was exactly as before, and each time my subconscious called a different message. Then I realized the cause, and the hair rose on my neck. Not three feet from me, the electroscope I had charged sat close to the idol. No one and nothing else was near it. Yet even as I watched, the gold leaf tilted slowly down toward the vertical, at ever-increasing speed. In a few more minutes, it hung completely limp.

I told myself that there must be some simple, rational explanation, something Illingworth had overlooked. But shortly afterward I started to hear sounds: a distant scuffling, intermittent, but definitely coming from within the Museum. I had just about convinced myself it was only rats (not a prospect I greatly relished) when there came a skittering sound and something quite large bumped my leg.

I started up with an involuntary cry, and a moment later felt as foolish as I ever have: it was merely a cat. Not even a sinister black cat, but a large and lazy-looking tabby. After allowing me to pet it for a minute or two, it jumped upon the table and curled itself up to sleep wrapped right around the idol.

The anticlimax was such—no idol looks very fearsome with a large tabby cat snoozing upon it—that I shortly started to have some difficulty staying awake. The light was too poor to read my book with any ease, and for a time I sat in thought with my eyes closed. I was aroused by a sound like the crack of doom: starting up, I found bright daylight streaming all about, and realized with some embarrassment that I had been fast asleep.

The crash could not really have been as loud as I thought, for the cat remained immobile just as I had last seen it. I rose and tried to stamp the cramp from my feet as I heard a clatter of mops and pails. Illingworth and Adams entered with the cleaning women.

"I trust you stayed alert," said Illingworth severely.

I avoided answering directly. "There has been no sign of any mischief. I apprehended only one intruder." I pointed to the cat.

"Ah, the museum staff will keep feeding the odd stray: I am not surprised one gets locked in occasionally. Down you come now." The Director put his hand to the cat, and pulled it back with a sound of disgust. I went forward and felt the animal. It was quite cold and stiff, and had evidently been dead for some time.

"It is extraordinary," I commented, "the idol certainly seems most ill-omened, but then I suppose stray cats are never the healthiest of animals."

"A dead cat is hardly relevant," said Illingworth acidly. "The creature can scarcely have sabotaged a photographic plate without leaving any sign upon its wrapping. A pity the miscreant did not return, but at any rate I will develop the plate, just to prove that untampered with, the quality remains perfect."

He took it into a small booth under the stairs which evidently served him as darkroom. I made small talk with the Director for a few seconds until we were interrupted by a hail from Sherlock Holmes.

"Good morning, Watson! I must confess I felt a trifle guilty about your lonely watch: nothing to report? I thought as much. Well, I purchased some pastries as I passed Glockstein's. Can I tempt you, Watson? Director? I believe I hear the hissing of a kettle somewhere: perhaps a trade for some tea, or even coffee, might be arranged."

We were tucking in to our makeshift breakfast when there came a cry from the adjoining cubicle. We rushed in: Illingworth was standing unharmed, but pointing a trembling finger at the plate, which lay flat in a shallow developing tank taking up most of the floor.

His horror was understandable. The plate showed the clear outline of a skeleton! At first I thought the image of one of the great saurians outside had somehow been transposed there, but then the truth dawned: I was looking at a human skeleton, albeit grossly deformed, distorted, and blurred.

Illingworth turned on me. "If this is some juvenile prank," he shrieked, waving a trembling finger, but was interrupted by Holmes.

"One moment, sir." He stared at the plate, and then at me so intently that I thought he must share Illingworth's quite mistaken suspicion. "Watson, bring that chair here and squat down. A little further forward. In which pocket is your revolver concealed. The left? Ah, I thought so!"

He pointed to the image in the tank. "Gentlemen, you have all heard of Roentgen's famous photographs, taken using light of a wavelength which he calls 'X rays' and to which the human body is translucent. I am sure you know, Watson, how the technique promises to revolutionize medicine, as it becomes possible to inspect bone fractures and other defects from outside the body. Now, suppose that there is a powerful source of some such invisible radiation in the museum. Between the source and your highly sensitive photographic plate sits Watson. There you see the chair legs; this blotch is the revolver, which being metal was quite opaque. The bones are somewhat blurred but not much—Watson, you must have kept a most conscientious vigil, scarcely moving from your post for most of the night!"

Now that he had pointed it out, we could all see the image clearly.

"But it takes a great deal of electrical energy to produce X rays. What can the source be?" protested the Director.

"As to that, gentlemen, some simple triangulation will show us," said Holmes cheerfully. "Come outside. We can see where the plate and the chair were set, the marks are just visible on the floor. Replace the chair. Watson, and seat yourself precisely as you were overnight. Now I shall just take a couple of measurements from the plate."

In seconds, he had sketched two converging chalk marks on the floor. At their intersection stood the idol on the table.

"Evidently we were wrong to dismiss the awe of this object as mere superstition," he said.

"But it is plainly solid metal, containing no mechanisms, and certainly no electrical apparatus. How could it possibly be emitting X rays?" Illingworth asked skeptically.

"Not necessarily X rays, but something penetrating, certainly," said Holmes. "My guess would be some form of electricity. We need something more convenient than a photographic plate to detect it unambiguously, and something sensitive enough to detect particles the size of a single atom."

"That will certainly require elaborate and exotic apparatus," I commented.

The Director smiled. "It can be done with air and water, using a simple technique I have devised. Fortunately, I have some small expertise in detecting invisible radiations." We watched rather baffled as he went to fetch a large syringe with a blank face of thin glass in place of a needle. He opened it and blew in some steam from the kettle which had recently made our coffee.

"When air expands, it cools," he explained. "Water vapor present forms droplets. The droplets do not form at random, but condense preferentially on any particles or electric charges present, however minute."

He held the syringe with its end close to the idol and pulled the piston out sharply. I gasped in astonishment. For just a second, before they evaporated, multiple trails of tiny cloud droplets were visible.

"It looks as if the idol is shooting streams of particles like a nest of machine guns!" I exclaimed.

"Not necessarily. The technique is so sensitive that each stream of bubbles can indicate the path of a single particle, as it disturbs consecutive atoms of air along its route."

"Astonishing, that such delicacy can be achieved so easily!" I cried.

The Director looked at me rather severely. "The essence of intelligent design is to use the principle of the lever in some more refined form," he said. "Be it in a mousetrap or an atomic detector, the right kind of trip-lever can always trigger an arbitrarily large effect. Perish the thought that a scientist should claim expensive and complex instruments are necessary to measure the phenomena of nature, when the key item is always the human brain."

"We await no doubt the telescope made from air and water," sneered Illingworth.

"We can confirm a suspicion of mine with the aid of this pocket magnet," said the Director. Holmes held the magnet close under his direction as he repeated the syringe action and smiled with satisfaction.

"Some trails curve very slightly clockwise, and some very strongly anticlockwise, in a given field strength," he remarked, "showing that the particles are of two types, one type negatively charged and one type positively charged."

"And the negative much more strongly charged, so it curves tighter." I said.

The Director nodded. "Or alternatively they have similar charge, but the positive ones are much heavier," he said. "At any rate, gentlemen, we must apologize for having troubled you. It is clearly a technical rather than a criminal matter."

"Do not mind Doctor Illingworth's manner," said Adams apologetically as he walked us toward the door. "He is a man who feels that anything of this mere Earthly sphere is almost too trivial to notice. But I think this chance finding may have brought some important new phenomenon to notice, which a humble chemist like myself will not be too proud to investigate."

A MONTH OR SO later, I came down to dinner to find Holmes reading a lengthy missive bearing the distinctive crest of the British Museum.

"Good evening, Watson. You will be glad to hear that if Doctor Illingworth has forgotten us, the Director is a little more remindful of his obligations. He has enclosed a check for our investigation. No, I insist you take it, Watson: but for you I should never have got involved in the case, and it was certainly your night's work that gave us the vital clue.

"He also thanks us effusively for drawing the idol's strange substance to his attention. He has been studying it further, with intriguing results. You recall the idol emitted two sorts of charged particles: the massive positively charged ones, which he now calls Alpha rays, and the lighter negatively charged ones, which he now calls Beta rays. He has also found associated rays which are not particles, which he calls—"

"Gamma rays?"

"Very good, Watson: you remember your Greek alphabet. In

any case, he has divined the nature of each. In atomic theory, it is strongly suspected that atoms are made up of charged particles, positive and negative, and that when an electric current flows, the lighter negative particles, called electrons, move while the heavier positive charges remain stationary, providing as it were a fixed supporting matrix.

"He has found that the heavy Alpha particles are simply atoms—that is, their positively charged parts—and specifically atoms of helium, having a double positive charge. The light Beta particles are electrons, each with a single negative charge. And the Gamma rays behave like light of very short wavelength, or in other words like X rays."

"So the so-called radiation from the statue is really made up of three rather dissimilar stuffs?" I asked.

"And quite possibly with more to follow: these may just be the commonest or most easily detectable sorts."

"And the radiations are harmful, it would seem?"

"In varying degree. The Alpha rays, being quite massive particles, do the most damage. Living tissue left exposed to them for a short time and examined under a powerful microscope is quite visibly smashed up."

A chill ran through me. "And I sat exposed to them for many hours!"

"Fortunately, no, Watson. By some lucky balance of nature, the Alpha rays, although deadly, cannot penetrate matter readily. Even a thin sheet of card, or a foot of air, will absorb a good proportion. They can have harmed only those who actually touched the idol, probably causing the strange burns and killing that unfortunate cat.

"The Beta rays are about one hundred times more penetrating than the Alpha, and the Gamma again one hundred times more penetrating than the Beta. The Director says that the Gamma rays can easily be detected at the far side of a foot of steel armor plating. But he says that allowing for the range at which you sat, and the duration, he is pretty certain no harm was done to you.

"By the way, the flow of charged particles makes the air near the idol into a conductor of electricity—not like a metal, of course, but just fractionally conducting—and that is why you saw the electroscope discharge."

"Well, I am glad Adams is grateful, Holmes: these sound like intriguing scientific findings," I said.

"Not at all, Watson, the exciting bit is yet to come. The Director reckons he will make his reputation from two really dramatic discoveries.

"First, he claims a most remarkable observation about the nature of matter. He was curious to investigate what would happen if he passed the Alphas through an exceptionally thin sheet of material. The actual substance used was gold foil: he chose that metal because it can easily be hammered down to a leaf just a few atoms thick, the same property which makes it ideal for the electroscope.

"I think he expected that the Alphas would penetrate but emerge with slightly reduced speed, as would happen if you fired revolver bullets through a blanket, for example. Instead he found something quite extraordinary. Most Alphas pass through the sheet as if it were not there at all, but a few are deflected through very large angles, some even reflected through a hundred and eighty degrees."

"That sounds rather surprising."

"Rather surprising, Watson! I suppose you would describe it as just a touch unusual, if you were to fire a fifteen-inch naval shell at a piece of tissue paper, and it rebounded and hit you on the nose. Do you not see what it implies?

"Let me explain by example. You no doubt notice upon the sideboard one of Mrs. Hudson's blancmange puddings, justly renowned for its lightness, and I hope destined for our dessert. Tell me what you would deduce if I borrowed your revolver and fired a shot into it at point-blank range."

"I would deduce, Holmes, that astonishingly tolerant of you as our landlady is, there is such a thing as a final straw, and that—"

"I had not finished, Watson. I was about to add: supposing the bullet, instead of penetrating, bounced back in the direction from which it came?"

"Then I would deduce that some humorist had replaced the pudding with an armor steel model, painted pink."

"And suppose you weighed the pudding, and found it weighed no more than it should? And if you fired many bullets, and found most penetrated without hindrance, with only a few rebounding?"

"Well, I suppose I should deduce the pudding contained just a few widely spaced dense objects—perhaps concealed gifts for the eaters, just as Scottish mothers traditionally hide pennies in the mashed turnip as surprise presents for their children at Halloween tea."

"Very good, Watson! But if the pudding really weighed no more than it should, you might have to conclude that the pink foam was scarcely more than illusion, and that almost all the mass was in the embedded coins. And you could deduce from the proportion of bullets that rebounded, and the angles, what part of the total volume was made up of coins, and the average mass of each coin."

"I daresay. But I still do not see what is so remarkable."

"Why, Watson, for a start it confirms the atomic theory: the hard embedded objects have just the weight expected for individual atoms. But it demonstrates something much more astonishing: that in a sense all matter is illusion! Even something as hard and dense as gold is much more than ninety-nine percent empty space. In fact, the Director estimates that the atoms make up just one part in ten to the fifteen—one thousandth millionth millionth—of the volume of any solid."

I flung back my head and laughed heartily. "Really, Holmes, I think it will soon be time for Mr. Wells to cede his crown! If what you say were so, for one thing I would be able to walk through solid walls with but the tiniest chance of any of my atoms colliding with those of the wall. The Invisible Man will have to make way for the Insubstantial Man. Let me see, can I put my hand through the table?"

I feigned surprise as I brought my hand down on the coaster, making the dinner things rattle.

"Nonsense, Watson; the electromagnetic forces between the atoms stop you. It is analogous to thinking you could attach powerful magnets to the prongs of two garden forks, and then insert the tines of one fork between the tines of the other without noticeable forces arising. It is just the combined force of all these millions of tiny atomic magnets, which repelled your fist just now, and will cause you a nasty bruise if you attempt any wall-walking feats.

"Practical applications will have to be a little more subtle, but even so, Watson, it is a most remarkable and unexpected insight into the nature of matter.

"The Director has also attempted a chemical analysis of the statue material, and discovered that although it is predominantly a metal, uranium, there are traces of other materials. He has isolated one chemical which seems to be responsible for a disproportionately high fraction of the Alpha and Beta emissions, and measured the energy generated. Each gram of this active substance, dubbed Radium, gives off enough energy to boil its own weight of water in an hour."

"You mean it is slowly combusting, no doubt with atmospheric oxygen?"

"No, the process continues hour after hour, even isolated in the absence of oxygen. The total energy given off is thousands of times more than any conceivable chemical reaction could produce. Moreover the only change he could detect in the sample, even after a period of some weeks, was a minuscule loss of weight. Although he did find on a second chemical analysis that the Radium seemed to have become contaminated with other elements."

I could not resist a little jibe.

"Well, the crackpots will have a field day with these revelations, Holmes! Not only has he discovered an inexhaustible source of energy, but the substance is also a sort of Philosopher's Stone: it can transform itself into other elements, such as helium gas. Rather than making the Director's reputation, it sounds to me more likely he will lose whatever reputation he already possesses. Surely to claim such unlikely results will merely bring disbelief and opprobrium, as to the sailor who reports the sighting of a sea serpent."

"Not so, Watson. The crucial difference is that the experiments can be checked and verified and repeated independently, as I am sure they will be when the statue is made available to other investigators."

"This makes me feel most uncomfortable, Holmes," I said. "I had just come to grasp how physical scientists understand the Universe in terms of certain quantities which are always rigorously conserved. Now you are telling me that, quite paradoxically, energy can come from nowhere, and elements change one into another."

Holmes rubbed his hands keenly.

"I have often observed to you, Watson, how important the exception which disproves the rule can be. It forces the investigator to recast his ideas from the beginning. It is precisely the appearance of such singular features in otherwise mundane circumstances which has led to my most intriguing cases, as those who follow my career through your narratives are aware.

"Until now, I have had little curiosity about the physical sciences. All seemed too cut and dried. Newton long ago inferred the laws governing masses and forces and motions, and in our lifetimes Maxwell has described exactly those which rule electrical phenomena. I have even heard it alleged that the work of future investigators might be confined to making ever more accurate determinations of physical constants already known.

"Now this knowledge will be put to a stern test. There is no such thing as a minor rewriting of physical laws. I would take a sizable bet that the coming months are going to be most interesting ones. Perhaps the matter may even turn out as important as Doctor Illingworth thought, albeit for quite different reasons."

He sighed, and stretched his arms wide.

"But let us put such cares aside for the time being. I can see that Mrs. Hudson is about to serve the blancmange, and even if it is in some fundamental sense illusory and mostly empty space, it will still be quite real enough to satisfy me."

5

The Case of the
Flying Bullets

"Ah, Watson, this promises some fun!" Sherlock Holmes held up a small buff envelope.

I felt considerable relief. For days, my colleague had been pacing the flat restlessly, hoping for some interesting problem to come along, and I had been increasingly nervous that his need for mental excitement might lead him back to the dark habit which I hoped he had put behind him for good. I was nevertheless surprised, for the envelope looked quite ordinary, and carried a less than exotic Central London postmark.

"Do you not recognize the handwriting, Watson? It is from brother Mycroft. It is rare for him to ask my help, but when he does, the case is always intriguing. And he is an economical man with words, yet the enclosure here feels to be several pages long: no trivial matter. The silver letter opener, if you please."

I passed the implement across. He slit the envelope and read the contents rapidly. Then his face fell in disappointment.

"Not the request you had hoped for, Holmes?"

"Hardly! He writes to complain that he is being pestered by two gentlemen whom he refers to rather inexactly as friends of mine."

"And who would they be?"

"None other than our scientific acquaintances, Professors Challenger and Summerlee. It appears they have approached him to sort out a scientific dispute between them."

"I hardly knew your brother was a scientific expert."

"He is not. Nevertheless, when he was at Cambridge, he was well known for his ability to help fellow undergraduates who were stuck on difficult points of their subjects, of whatever discipline. He has a remarkable faculty for bringing pure logic to bear on any problem, however thorny, and making the resolution seem a matter of plain common sense. His contemporaries at university remember his talent, and from time to time he still gets requests such as this."

"Even in the sciences? I had remembered him as a historian and linguist."

"Especially in the sciences, Watson. He confided in me the secret of his technique. He calls it performing thought-experiments."

"I can hardly imagine your brother at a laboratory bench!"

"Nor I, Watson. Physical work and application are scarcely Mycroft's strengths. But he is often able, by the sheer power of his mind, to devise an imaginary experiment whose result can be easily inferred, and which throws light on a problem."

He tossed the letter down on the table. I could see that it bore the letterhead of the Diogenes Club, that remarkable gathering place for recluses on whose premises ordinary conversation was forbidden.

"He says that the two Professors have become engaged in an increasingly bad-tempered debate on the nature of light. Challenger holds that light takes the form of a wave, spreading continuously through space from its source like a ripple on a pond, whereas Summerlee believes it to be composed of a stream of tiny particles."

"This reminds me rather of the debate about atoms, Holmes. Just as atoms have won out against continuous matter, and germs against the miasma theory of disease, so let us go by precedent and assume light also is composed of discrete units."

"Reasoning by analogy is very suspect, Watson. Analogy may be fruitful as a source of ideas, but it can never provide proof."

"Still, I can readily imagine light as a jet of tiny corpuscles," I said. "Sprayed at great speed, so to all intents they travel in straight lines, and bouncing off mirrors just as an indiarubber ball bounces off a smooth pavement."

"That is what Summerlee maintains, Watson. And he may be right. But I understand there is also a persuasive case for the wave view. For example, it is known that light travels more slowly through glass than through air, and also that light bends inward as it passes from air to glass, and back outward as it passes from glass to air. Have you ever seen an ocean wave pass over a submerged reef? The wave speed over the reef reduces, and that bends the wave front, in a way which is quite natural when you witness it. So the wave theory neatly explains the functioning of lenses and prisms and suchlike devices.

"The wave theory also explains the existence of different colours of light: they are just waves of different lengths. For example red light has a longer wavelength than green, and green than blue. And the notion extends to radiation beyond the visible. An object which is not hot enough to glow red radiates heat at infra-red wavelengths, which are longer than visible red, and the X rays your medical colleagues are putting to such wonderful use have wavelengths much shorter than the visible blue.

"But the most convincing evidence that light is a wave comes from the work of the great James Clerk Maxwell. He posited that radiation of the same kind as light could be produced directly by electricity. We now know these rays as Marconi, or radio, waves. They are rays whose presumed wavelength is much longer even than infra-red, and—"

"That was a phenomenon worth mastering!" I interrupted. "I was reading the other day how, if every ship that sails were equipped with Marconi apparatus, the loss of life in shipwrecks could be greatly reduced, as a message for help could be transmitted to other vessels, even beyond the horizon."

"Do not believe everything you read in the popular scientific press, Watson! If even one-tenth of those bright ideas published could be brought to fruition, the world would be transformed. But where was I? Ah, yes: Maxwell showed that oscillating an electric charge is just the mechanism that causes light waves to be produced. I can demonstrate, Watson, if you will ring for tea."

I was somewhat surprised by this abrupt request, but I did so, and a few minutes later Mrs Hudson's eldest daughter—Mrs. Hudson herself having taken a sudden and somewhat mysterious holiday—brought in the tray. I poured for us both, adding a little milk which made the liquid opaque, and Holmes picked up a teaspoon.

"Observe, Watson, if I take the blunt end of the spoon, and I wiggle it up and down in the tea, so—"

"It sets up a ripple."

"Quite so. And if I change the frequency of the wiggle—"

"The wavelength of the ripple changes."

"Brilliantly observed, Watson! And similarly, Maxwell showed that the emission and absorption of light is caused by the vibration of electric charge, just as wiggling the teaspoon causes ripples in your tea. Wiggle the charge faster, and the wavelength becomes shorter. The wave is detectable by electrical instruments as an oscillation of electric and magnetic fields."

"Say no more, Holmes, I am convinced: light is indeed a wave."

I sat back and sipped my tea. A moment later I sat bolt upright.

"Holmes, I have just had an insight. On the contrary, the wave theory is all nonsense!" I felt quite excited. Medical men have often made valuable scientific contributions, but I had never dreamed I should be among them.

"Consider, Holmes, light travels through substances that are transparent. Commonly glass, water, and air."

"But not through granite or Stilton cheese, for example. Impeccably observed, my dear fellow."

I was too excited to be offended by his tone. "But there is something else through which light travels, Holmes. And that is— nothing! For we know that the atmosphere gives way to the void of space a few tens of miles above our heads, and yet we can see the stars. Now water waves require water, and sound waves require air—"

"Not necessarily, Watson. Sound waves can also travel through solids, even such as bricks and mortar. Our neighbors have remarked on this in connection with late-night violin playing, as you remind me occasionally."

"Yes, but water, air, bricks: all constitute matter. Any wave is a form of motion, and so requires something to move, or it has no existence. Light can travel through a vacuum. So it is no mere wave."

I sat back triumphantly. "If I pen a letter to the journal *Nature*, Holmes, would you be so good as to countersign—"

Holmes smiled and held up a hand. "Just one moment, Watson. You are quite correct that light can travel through vacuum, and fortunately so. Else we should be deprived not merely of the stars but of sunlight also.

"But specialist scientists are not entirely fools. This point has in fact occurred to them. The problem is solved by postulating that the universe is uniformly filled with ether."

"Ether, the anesthetic gas?"

"No, ether denotes an intangible substance, filling every corner of space"—Holmes flung his arms wide—"which supports the propagation of electromagnetic waves. You could say that ether is to a gas, as gas is to a solid. It flows through the densest object, even the Earth itself, as easily as air through a butterfly net, and so is imperceptible to our senses in all ways but one: its ability to transport those waves of electrical and magnetic force which we discern as light."

"It seems a rather large assumption, Holmes, to fill the universe with invisible and intangible ether just to explain the propagation of light rays," I protested.

"There are plenty of things which exist and are indisputably real, despite not being perceptible to your unaided senses. The ether could turn out to be as real as the Earth's magnetic field."

I took a sip of tea, restraining myself from any profound thoughts about the ripples created in the cup.

"Well, it is beyond me, Holmes. But Mycroft expects you to solve the dilemma?"

"No, Watson. He says the evidence is contradictory and confusing, and he does not believe a solution will be found soon. He has a more humble role for me. It appears that Challenger and Summerlee have offered, or from Mycroft's point of view threatened, to come and debate their points of view before him, acting as arbiter."

"Good Lord! That can hardly be to Mycroft's taste."

"No indeed. As you know, he prefers to avoid his fellow man as far as possible, and he usually declines even the least demanding of social invitations. But never fear, his great mind has found a solution: he wishes to foist the unpleasant task on his long-suffering younger brother. He puts it a little more tactfully, of course."

He rose from his chair and paced restlessly.

"This is disappointing, Watson. I shall expend some tiny intellectual effort in composing a plausible excuse: it will do Mycroft no harm to confront the fact that even his mind runs up against the occasional problem of Man or Nature which is beyond it. But ah, how I need some more engaging puzzle to occupy my mind today. A client, a client, Watson, my kingdom for a client!"

At that very moment, the doorbell pealed. Sherlock Holmes strode to the window.

"It looks as if my prayer is answered—ah, confound it!" He sprang back from the window, hastily gathered the tobacco-tin into his dressing-gown pocket, and strode to the door of his bedroom.

"Just remember that I am out, Watson, and it is quite uncertain when I shall be back."

I looked at him in bewilderment.

"But you were just saying—"

"I was referring to puzzles of a significant sort, Watson. Heaven knows what he wants today. Probably a test tube has gone missing, or a textbook been misplaced, and he expects a major investigation."

At that moment there came footsteps on the landing, and Sherlock Holmes placed a finger to his lips and quietly closed the inner door behind him.

I rose to greet the visitor, and the motive for Holmes's extraordinary behavior became clear: it was none other than the obsessive savant of our recent acquaintance, Doctor Illingworth. An expression of mingled triumph and fury was upon his face.

"Good morning, Doctor. I am seeking your colleague Sherlock Holmes. The most sinister event has just occurred. This will teach him to dismiss my just fears of scientific sabotage as mere trivial fantasies."

I sought to calm him. "Pray have a seat, sir. I am sure whatever has befallen you is most upsetting. Has your work been seriously inconvenienced?"

Illingworth snorted. "It is not a question of mere inconvenience, Doctor. A young life is wasted—tragically wasted. If only your colleague had listened to me, the poor man might yet be alive."

I was able to infer that some fatal laboratory mishap had occurred, which in his paranoid fashion Doctor Illingworth had attributed to some enemy. I decided to demonstrate that I also had some deductive powers.

"Any lethal accident is tragic," I said in my most soothing manner. "But even the best-regulated laboratory is inevitably a dangerous place, and we are all fallible—"

Illingworth snorted. "Accident? Laboratory? What are you blathering about, Doctor? I am speaking of a man gunned down, shot from afar with a high-power rifle while outdoors, not some error at the chemistry bench."

I saw that I had jumped to conclusions.

"I do beg your pardon, sir. Unfortunately, Sherlock Holmes is away just at present; I do not even know if he will be back today. I will of course be glad to pass on an urgent message, but I fear it would be pointless waiting—"

At that moment the bedroom door opened, and Sherlock Holmes strode into the room.

"What? Not here? Really, Watson, you must clear the fug from your brain. To think you had not noticed my return! Now, Doctor Illingworth, I am never too busy to investigate a murder. Let Watson take your coat—thank you kindly, Watson—and pray tell me exactly what has occurred."

Somewhat put out, I attended to this chore, as Illingworth launched into his narrative.

"A most important research project of mine—the same I was embarking on when we last met—has been proceeding at Runnymede Hall. Is the name familiar to you?"

"Ancestral home of that family, bequeathed to Cambridge University for the furtherance of astronomical studies by the last Lord Runnymede on his death eight years ago."

Illingworth looked surprised. "I see you are not quite ignorant of matters scientific. As you say, the place is now an observatory, and seemed quite ideal for my project. My time is of course too valuable to spend on practical observation, and I sought research students to work on the task.

"For some time I had difficulty finding volunteers, despite the priceless benefit of personal contact with myself that the project would entail. Then quite unexpectedly a young woman came forward.

"She had no academic qualifications of any kind, but most exceptionally the faculty had admitted her for doctoral study. Apparently she had made some name for herself as an amateur astronomer, having made several significant discoveries, no doubt by pure good fortune."

He snorted. "It is my considered opinion that the attempts of those without formal qualifications to contribute to science are of little value, and moreover that the proper place of a woman is in the kitchen. However, in the absence of other volunteers, I was forced to accept her. Strangely enough, no sooner was this announced than two young men also came forward."

"Strange indeed! She is attractive, this young woman?" asked Holmes.

"I would judge so, although I do not perceive the relevance. To make maximum use of the observatory, it was arranged that each of them would do a full night's duty in turn—stand one watch in three, so to speak. In fact, it soon came to my attention that on Miss Latham's night, one or both of the young men, Tom Phipps or Martin Hennings, would often volunteer to assist Mary, no doubt because they wished to ensure that her lack of professional training did not lead to lapses."

"No doubt," said Holmes dryly.

"The three students continued to reside at their Cambridge addresses, the two young men in their colleges, as required by university statutes. Fortunately Runnymede Hall is close to the village of Shelford, which is on the main London to Fenland railway line. During the day trains stop at the village station—at night only expresses run direct from London to Cambridge—and so the three were able to travel to and from their duties easily.

"For the first weeks, all proceeded smoothly. Hennings and Miss Latham performed their duties admirably. Phipps also performed well, but spoiled his record by failing to report for duty on two occasions. Each time, he claimed to have narrowly missed the six o'clock train from Cambridge. The later expresses do not stop at all between Cambridge and London, and Runnymede is too far from Cambridge to reach by other means, so I had to accept the excuse."

Holmes nodded thoughtfully. "What more can you tell me about this Phipps?"

"Well, he is a couple of years older than his contemporaries. Apparently he was in some quite serious trouble while still a very young man at public school. He was involved in a duel, fortunately without fatal consequences to either party, and his family sent him to relatives in India. He did well there, putting his marksmanship to better use by becoming expert in tiger shooting. For safety, tiger hunts are normally conducted riding on elephants, and he became known as a crack shot from this vantage, disposing of several man-eaters, to the relief of the local population.

"After a year of good reports, his family permitted him to return. He had after all been very young at the time of his indiscretion. And the proctors, making allowance for the same reason, decided that he was now a fit person to be admitted to the university.

"I gather the decision has barely justified itself. He is an intelligent young man, but has been in several minor scrapes with the Bulldogs, although nothing serious enough to warrant sending him down."

I must have showed my bafflement, for Holmes explained: "The Bulldogs, Watson, constitute the University's own police force. They are recognizable by their bowler hats, and have a reputation for holding tenaciously to any young man they apprehend, hence the nickname. But," he turned back to our visitor, "pray continue your most pertinent account."

Illingworth shook his head. "Quite frankly, sir, it is not pertinent at all, for although Phipps is certainly the kind of young man who may have made enemies, he was not even present last night when the tragedy occurred. Miss Latham was on duty, assisted by Hennings. According to Miss Latham, having set up the photographic plates in the observatory dome, they stepped outside. The dome is set atop a folly immediately adjacent to the house, and there is a good view over the countryside from the balustraded walk which surrounds it."

"Scarcely such a good view in the middle of the night."

"Well, that is the explanation Miss Latham gave. As they happened to be standing close together, it appears that several shots were fired, although Inspector Lestrade seems to think her account must be imperfect, shock no doubt having confused her memory."

"Ah, so the good Inspector Lestrade is upon the scene. I assume you are less than happy with his progress, to have come to me?"

"Quite so. He seems to be obsessed with some romantic notion he has concocted, of love and eternal triangles, and is quite unable to comprehend that the matter is likely to be a deeper and more sinister one of scientific rivalries and sabotage."

To my amazement, Holmes nodded. "Indeed, I quite agree the matter needs looking into at once. Are you free to accompany us, Watson? Splendid! If you would like to descend, Doctor Illingworth, we will be with you momentarily."

I looked at Holmes in some bafflement as the door closed behind Illingworth.

"Really, Holmes, this case is hardly worthy of you. Even I can see the solution, as evidently can Lestrade. Do you place no higher value on your time?"

Holmes smiled. "I daresay you are right, Watson, although you should never leap to conclusions in advance of the evidence. But my reasons are more strategic. Mycroft implied in his letter that he might proceed to call on me, in which case Challenger and Summerlee will assuredly be waiting in the wings. Given a choice between a day listening to two bad-tempered scientists debate some obscure point of physics in a stuffy room, and one in the countryside investigating a case in the company of your good self, I have no hesitation. And if all is as straightforward as it sounds, for once I may be able to agree with Lestrade, which will help my relations with the Yard not a little. Come, Watson: let us go to the challenge of Runnymede!"

We disembarked from the train at the Shelford halt, and hired the waiting trap to take us the two miles or so to Runnymede. The horse clopped at a leisurely pace through the flat Fenland countryside, Illingworth fretting while my friend gazed about calmly. I felt some concern myself, for storm clouds visible on the horizon were blowing rapidly toward us, and if we did not reach shelter in time we should evidently be soaked. At that moment, a flash of lightning lit the horizon. Some ten seconds later, we heard the clap of thunder.

I sought to impress my colleagues by a scientific remark. "One can tell the distance of a storm by the delay between the lightning and its associated thunder. Sound travels at three hundred and thirty meters per second, so that one must be three and one third kilometers, or just two miles, away. We had best make haste!" I said.

Illingworth looked at me disapprovingly, as at a student who has made some observation too banal to comment on.

"Of course, that formula implies that the light from the flash itself travels infinitely fast," I said. "For example, if light traveled only ten times as fast as sound, I would have underestimated the distance by one-tenth."

Illingworth snorted. "In principle you are of course correct, Doctor. But in fact the speed of light is about one million times greater than that of sound, three hundred thousand kilometers per second, so I would hardly concern yourself with the correction."

"Why, that is quite amazingly rapid," I remarked, determined to keep the peace. "In fact so rapid, I am astonished anyone should have noticed it. After all, one is normally quite unaware even of the limited speed of sound."

Illingworth frowned but made no reply. We sat in silence as the trap entered the estate grounds. We disembarked in the drive-way just as the first drops of rain started to fall, and were admitted by a housekeeper into an imposing front hall, where Lestrade was standing in coat and hat, accompanied by a young woman with long dark hair and strong features. Lestrade appeared unsurprised to see us.

"Good day, Mr. Holmes. And to you, Doctor Watson. Doctor Illingworth intimated he would like a second opinion. But I fear there is no longer any mystery to clear up. I have just been examin-ing the evidence that really clinches the matter."

He held out his hand. In his open palm were two spent bullets, subtly different in shape from anything I was familiar with.

My friend examined them keenly. "German manufacture—undoubtedly ammunition for a Macher gun," he pronounced. "Quite a specialist item. From the same batch of manufacture, and fired from the same piece."

"And Phipps was by way of being a professional shooter, I understand," said the Inspector cheerfully. "Now I must love you and leave you, for I am on my way to check out an alibi he claims in Cambridge. I have no doubt I will be able to disprove it, for the circumstantial evidence points unmistakably to him."

He walked toward the door with us, and said in a lower voice, "You are welcome to talk to Miss Latham here, but be warned the shock may have slightly affected her memory." He raised his hat to us and was gone.

Illingworth indicated the young lady. "Mary Latham, allow me to introduce Sherlock Holmes, Doctor Watson. Please describe the events of the night to them. I must secure the telescope, before this rain gets into the dome."

He bustled off. The girl, who appeared very calm and com-posed, showed us through to a small side parlor, and bade us to be seated.

"There is really not much to tell, Mr. Holmes. You know that Martin, Tom, and myself made up the observing team here," she said.

Holmes nodded. "I see that you were on first name terms," he observed.

"Whereas Doctor Illingworth must be a difficult man to get along with," I said. "I tried to draw him out in a scientific conversation, on the subject of the speed of light, and received the impression he cannot be an easy tutor."

Miss Latham smiled. "That was an unfortunate choice of subject. Some time ago, he decided to make some observations of the moons of Jupiter. He wished to time the occultation of one of the smaller moons recently discovered—the moment when it passes into Jupiter's shadow—to verify if its published orbital parameters were accurate. An occultation can be timed to an accuracy of a few seconds, so it was a good test of previous measurements.

"To his gratification, he found an error of over ten minutes relative to the calculated time, and immediately wrote a letter to *Nature* denouncing observational astronomers as incompetent fools.

"Unfortunately, he had overlooked the fact that at the time of his own measurements, the Earth was some two hundred million kilometers farther from Jupiter than at the time of the published observations. Light takes some ten minutes to cover this distance, which explained the delay: the original measurements had in fact been perfectly accurate.

"What was particularly embarrassing for him was the fact that, many years ago, it was just such a discrepancy, in precisely similar circumstances, that originally caused scientists to notice the speed of light was finite, and to make the first quantitative estimate of it."

Holmes smiled. "Thank you—you have cleared up a small mystery for my colleague. But at any rate, you three research students were on friendly terms?"

"Yes. Indeed there is not much point in concealing from you, Mr. Holmes, that both men showed signs that they would have welcomed some relationship closer than friendship with me."

Holmes smiled. "I can quite believe it. Did this not make for some problems in your work with them?"

"Not really. I made it clear at the outset that I valued both as colleagues, but sought no advances. In fact I liked Martin rather better than Tom, but I took care to conceal the fact, in the interests of good teamwork. That is, until recently—" She stopped abruptly.

Holmes nodded gently. "Please tell us about the events of last night," he said.

Her manner hardened. "I do hope you will listen carefully, Mr. Holmes. It is clear the Inspector thinks my memory must be at fault. This is quite galling to me, for I am a trained observer, and I recall the events with dreadful vividness.

"Martin and I had set up the plates for the first observation of the night. But it was still only dusk, and we went outside to wait for full darkness on the roof of the folly. We must have been very visible upon the parapet. We were standing close together—very close—"

"You were embracing?"

She reddened. "Yes. I must tell you, for it is pertinent. Suddenly I felt a shock go through his body. An instant later, I heard the crack of a rifle.

"We were spun half around by the impact. A second or two later came a second rifle crack, immediately followed by the shock of a second bullet."

"You could not tell from where the bullets were coming?"

"No. They could have been fired from anywhere in the grounds. We must have been very conspicuous, upon the rooftop."

She looked to be in some confusion. "I am really forgetting my manners, gentlemen. I suppose I am the hostess here. Can I not offer you some tea? No, please do not get up; I will be back shortly."

As the door closed behind her, Holmes smiled tigerishly.

"What do you make of her story, Watson?"

"Well, it seems straightforward enough."

Holmes shook his head. "It is the sequence of events that I find baffling, Watson. First comes the impact of a bullet; afterward she hears the crack of the rifle."

"That is easily explained, Holmes." I shuddered: my memories of sniper fire in Afghanistan still return to haunt me sometimes. "There exist rifles which fire bullets faster than the speed of sound. The bullet itself thus reaches you before the noise of its firing. They are deadly, Holmes: the lack of a split second's warning to duck makes them that much more dangerous."

"I am well aware of the devices, Watson. But recall the young lady's description of the second shot."

I paused to think. "Why, in that case, the sound did precede the bullet. So the second shot was slower than sound."

"Very singular, Watson. The velocities differed, and the first shot must have been going at least one-tenth faster than the speed of sound, the second at least one-tenth slower, or the delays would not have been perceptible."

"Perhaps if the ammunition was of very poor and variable quality—" I ventured.

"Unlikely, Watson. The Macher does not take standard cartridges, but uses its own proprietary ammunition. The latter is known for its high quality, and in fact is carefully crafted to make the bullet leave the muzzle at precisely the speed of sound. Thus it neither gives advance warning to the target, nor is wasteful of energy: it is shaped so aerodynamically as to lose hardly any speed in flight. Hardly consistent with what Mary has told us. Yet I believe her account, although I can quite understand why the Inspector does not wish to."

"Ah, you have picked up the point which he was too stubborn to grasp!" Illingworth stood in the doorway. "You see the implication?"

"I believe so," I said. "Different weapons, with different muzzle speeds, must somehow have been involved. Perhaps there are further bullets which have not yet been found. A jealous lover certainly would be acting on his own. Two or more men, with different weapons, implies an assassination."

Illingworth nodded, and ducked off down the corridor as Mary returned with a tea tray.

"So this could imply that Illingworth's theory of scientific sabotage, fantastic though it sounds, may be right," I said. "If he is really on the verge of a discovery of phenomenal importance, then who knows what measures some rival might take."

"Oh, I hope so!" Mary exclaimed. "I would like to think Tom innocent: he has a temper, but I am sure he is not a bad person at heart." She shook her head. "But what am I saying? There is the clearest evidence he cannot have been here at dusk yesterday."

Holmes looked at her thoughtfully. "You are referring to the alibi which Lestrade is now on his way to confirm?"

"Yes, and it is easily verified. You know that Cambridge under-graduates are required to be on college premises by nightfall, except in exceptional circumstances?

"Yesterday, at about ten o'clock in the evening, the porter on lodge duty at Old College was alerted by a knock on the mortice gate. He opened it, and a figure whose face was obscured by a muffler pushed past him and ran into the quad. The porter gave chase, assisted by two Bulldogs who happened to be in the lodge at the time. They soon caught the man, and unmasked him: it was Tom Phipps."

"Breaking into his own college?" I asked.

"Yes. It is a common tactic for young men who have been drinking in the town pubs, and wish to avoid having their names taken and being reported to the Dean for late return," explained Mary. "There was beer on his breath. The offense is not viewed very seriously, and the college servants permitted him to return to his room after taking his name.

"It is a mildly sordid story. But it gives Tom an unimpeachable alibi. The staff know him well, so could not be mistaken in their identification, and he is not popular with them, so they would certainly not be lying on his behalf. As trains from London do not stop here at Shelford after six o'clock, there is no way Tom could have committed the crime and been back in Cambridge so soon after."

Sherlock Holmes frowned thoughtfully. I noticed that he did not meet Mary's eye.

"This certainly complicates matters," he said. "I wonder if there is a smoking-room here, where I could pace about while I enjoy a pipe. You can amuse yourself for an hour or so, Watson? I will see you both in due course."

I passed the time by strolling along the path which ran around the house, the rainstorm having passed as quickly as it had come. The house proved something of a disappointment from an architectural point of view: the nearby folly did not match the design of the main building, and the metallic observatory dome atop it contrasted bizarrely with the ancient stone. However, I was presently joined by Miss Latham, and we made small talk pleasantly enough.

As we passed an imposing set of French doors, they were flung open and Sherlock Holmes emerged. "Miss Latham! Tell me, where exactly is the railway line which passes through the grounds?"

I looked at him in astonishment. The land about was completely flat, and a badger track would have been visible for miles, much less a railway line. But the girl merely nodded. She led us some fifty paces, to a spot near the folly. Invisible from more than a few meters away was a deep but narrow cutting, almost a tunnel, and within it two sets of rails.

"Few guess its presence, Mr. Holmes. A condition of running it through the grounds was that it be made quite invisible and inaudible from the house. In fact, even from the top of the folly, you can barely see the roofs of the carriages as a train passes. You must have been told it was here."

"Not at all: I deduced it. It was the only explanation that would fit the facts," said Holmes confidently. "Consider. A gun is known to fire bullets at precisely three hundred and thirty meters per second. The speed of sound waves in air is also three hundred and thirty meters per second.

"The gun fires two shots. Each bullet is fired from the gun at the same speed. The first travels faster than its own sound, the second slower. Where is the gun sited?"

Mary and I both looked at him in bafflement.

"Do you not see? The only possibility is that *the gun itself was in motion when the shots were fired.* I shall now tell you precisely how the crime was done.

"Yesterday afternoon, Phipps travels up to London. At a specialist gunsmith, he purchases the Macher, very likely using a license document with a false name.

"He buys a ticket for the eight o'clock non-stop from Liverpool Street station to Cambridge, selecting an empty compartment. At some point, he clambers out of the window, up on to the roof of the train. No great feat for an athletic young man.

"He settles himself on the roof. As the train passes into the cutting through the grounds here, he has a clear view of the top of the folly. He knows you will likely be outside the observatory dome: he has chosen a time when it is still light enough for him to see to aim, not dark enough for your observing to begin.

"He knows that his best chance will not be as the train passes closest to the folly: the angle will be changing too rapidly there for even the best marksman to take an accurate shot. The opportunity will be a second or two earlier, when the target is more nearly dead ahead of the train.

"He fires. The bullet leaves the muzzle at three hundred and thirty meters per second, *plus the speed of the train*—some thirty meters per second additional. The bullet travels comfortably ahead of the sound of the shot."

"One moment, Holmes," I interrupted. "Surely the sound wave would also be fired forward at a faster rate, so to speak."

"Not so, Watson. The speed of sound, or any other wave, is relative not to the speed of the *source* but to that of the *medium* through which it passes. A following wind can speed sound relative to the ground, but the speed of the emitter, be it a gun or a trombone, is quite irrelevant."

"I see! And the second shot?"

"The second shot was similarly fired as the train sped away from the tower. The bullet was correspondingly retarded.

"Of course, the alibi follows immediately. Phipps had only to climb back into the compartment, make a small fracas to ensure the time of reentering his college would be logged, and it was apparently impossible for him to have reached home from Runnymede so fast."

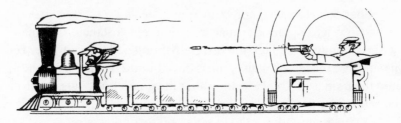

Shooting the Driver

"THE WONDERFUL THING is, Watson," said Holmes the following morning, as the express from Cambridge carried us back Londonward, "that the case has also given me the solution to Mycroft's little problem."

He rubbed his hands. I was aware that his opportunities to get one up on his almost supernaturally able brother were few, and correspondingly cherished.

"You see, Watson, if light is a particle, then it will travel at a speed determined by the emitter. If on the other hand it is really a wave in the ether, then—"

"Its speed will depend only on the motion of the ether."

"Exactly so. Let me explain it in terms of trains." He drew the sketch which I reproduce above. "Let us suppose that a train consists of flatbed trucks, except for the engine and guard's van. The guard wishes for some reason to assassinate the driver.

"Suppose that he fires when the train is stationary, and that the bullet takes one second to travel the length. The driver has exactly one second to live, at the moment the gun is fired."

"Very well."

"Now suppose that, desiring to delay his fate, the driver has brought the train up to speed, and then the guard fires. Is the delay increased?

"Not at all. In effect the driver is fleeing, but the gun and hence the bullet are traveling faster by the same amount. The velocity of the bullet relative to the train is as before, and the driver again dies after exactly one second."

"A bit hazardous for the guard as well, now aboard a train with a dead man at the controls," I said.

Sherlock Holmes ignored my quip.

"Now let us suppose that in a different instance, the guard wishes merely to frighten the driver, by firing a blank cartridge. Let us suppose the length of the train is three hundred and thirty meters, and the train is at rest. How long after the guard fires will the driver be startled by the bang?"

"Why, that works out to exactly one second, again."

"And now suppose the train in motion at high speed, the guard fires—how long this time?"

"Ah, I see! The train is traveling rapidly through the atmosphere: the latter, which conducts the sound, is still stationary, so it will be a little over one second before the driver is startled."

"Very good, Watson! And so the difference between a bullet and a sound wave is established. Now let us take the same train, and mount a photographer's flash gun in the guard's van. We are going to time precisely how long it takes for the flash to reach the driver.

"We do the experiment twice, once stationary and once at full speed. If the time is exactly the same in both instances, light is particulate, the particles fired from the lamp like bullets from a gun. If the time is just a fraction longer in the second instance, light is a wave in the ether."

"Why, that is quite brilliant, Holmes!"

Sherlock Holmes smiled tigerishly. "Just for once, I may surprise Mycroft a tad. In fact," he said as the train started to slow for the London terminus, "I will call upon him en route to Baker Street. I shall see you later, no doubt."

ENTERING OUR ROOMS some thirty minutes later, I was astonished to see Mycroft installed in one of our armchairs.

"Good morning, Doctor! Your landlady was kind enough to grant me entrance to await my brother. He is not with you?"

"No—ironically enough, I think he has gone to your club."

"Ah! Well, that may turn out for the best. The truth of the matter is, I came here partly because I had an intimation that Professors Challenger and Summerlee might turn up at the Diogenes, and I naturally wished to avoid them."

"It may well be for the best. Your brother has explained to me a simple way to find the solution of the wave-particle controversy, and no doubt will be glad to enlighten them."

Mycroft's eyebrows shot up.

"Really, Doctor? Well, I am certainly most indebted. Even more so, if you feel you might be able to explain the solution to me."

I fetched pencil and paper from the writing desk, reproduced Sherlock's train diagram, and explained the procedure as best as I could. To my astonishment, Mycroft's reaction was to sit back in his chair, and rock gently with silent laughter. He raised a hand at the sight of my expression.

"No, please do not be offended, Doctor! You have explained admirably, and it is amusing that Sherlock should have spotted this solution, and think it original. Alas, it has already been tried."

"What, using actual trains?"

"Not literally. Trains move so slowly compared with light— about one ten-millionth of its speed!—that the fineness of timing required would be quite beyond present-day instruments. But ways have been found to do the same experiments, making use of facilities provided by Nature. If I may borrow your pencil?"

He took a fresh sheet of paper, and sketched the diagram overleaf.

"Many stars visible in the heavens, Doctor, are in fact double stars. The two are usually greatly unequal in size, so the smaller effectively circles the larger just as the Earth orbits the Sun. With modern telescopes, we can measure the motion accurately, even though such star systems are so distant that the light itself takes some years to reach us. We can verify that the motion obeys Newton's well-known laws.

"The smaller star typically orbits at some tens of kilometers per second. Now suppose light was particle-like. On the side of the orbit where it is traveling away from us, the light would be approaching us correspondingly more slowly, by about one part in ten thousand. From the other side of the orbit, however, emitted light should be speeded toward us by a matching amount.

"Now, during its years of travel, the small difference in the

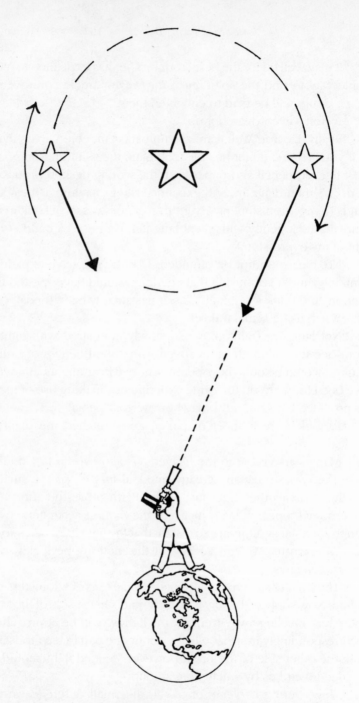

Light Shot from an Orbiting Star

supposed speed of light from opposite sides of the orbit adds up to a great difference in the distance traveled. We should see the light from the 'toward' side well early, and the light from the 'away' side well late. In fact, the *apparent* motion we saw should look nothing like a Newtonian orbit."

"Ah! Then you have neatly ruled out the particle hypothesis. Light must be a wave," I said.

Mycroft nodded. "The inference had been widely accepted, until a more recent experiment. You know that the Earth circles the Sun at some thirty kilometers per second. Let us perform an experiment to measure the speed of light in a particular direction at some point during the year—say, in the Spring. We will repeat our experiment six months later in the Autumn."

He completed a second diagram, as shown below.

"Now, while we do not know the velocity, if any, with which the Sun moves through the ether, the velocity of the Earth through the ether must certainly change, by a not negligible sixty kilometers per second, as it moves around its orbit.

"The speed of light through our apparatus should vary as the speed of the relative ether current varies. Our two experiments should accordingly yield slightly different measurements, and so the existence and direction of motion of the ether will be revealed."

I must have looked puzzled.

The Ether Sea: With and Against the Current

"Really, Doctor, it is the same as Sherlock's train experiment, just a little more cumbersome because we cannot speed up and slow down our planet quite as conveniently as a train."

"And what results were actually found?" I asked.

"The speed of light appeared identical in each measurement."

I felt a headache starting. "But then there is no ether current effect, and so having told me a few moments ago that light does not behave as a particle, you are now telling me that it is not an ether wave either. So what is its nature, then?"

Mycroft smiled. "You begin to see the problem, Doctor. There is just one solution that anyone has been able to think of: that it is indeed an ether wave, but the ether, being rather insubstantial, is dragged along with the motion of any matter it permeates, such as the Earth and its atmosphere. So no ether drag effects are perceptible to us. But in fact it seems most implausible the ether should do this. For one thing, any such dragging or swirling of the ether would change the direction of passing light waves, so we should see the constellations distort as the Earth turned, rather as the landscape seems to ripple when you watch it pass by through a railway carriage window of old-fashioned glass whose surface is not perfectly flat. No such anomalies are observed."

Mycroft seemed to be growing almost comically earnest. I really felt unable to take the matter so seriously.

"I am sure some solution will be found in due course, as more refined experiments are performed," I said.

Mycroft shook his head. "No, Doctor, the problem is not one of inadequate data, but inadequate understanding: the apparent impossibility of interpreting that data we already have."

I tried to think of some intelligent remark to make, but quite frankly I was completely floored. I tried to conceive of some other topic of conversation, but felt most inhibited: no doubt from his intellectual pinnacle, Mycroft would find any small talk of mine banal indeed. Then my eye lit on that morning's paper. On the back page, beneath the crossword, were printed a couple of brain-teasers: logical conundrums with which I had been passing the time on the train. I had found them utterly baffling: perhaps Mycroft would derive a moment's amusement from their solution. And if he too was stumped, it would be no small consolation to my ego.

I picked up the paper. "I say, what do you make of this? A young man is run down by a brewer's dray. Severely wounded, he is put aboard the cart and rushed to a nearby hospital, where he is at once taken into the operating room.

"The chief surgeon enters, picks up the scalpel, looks down, then gasps in horror: 'I cannot operate on this man. He is my son!' But now here is the puzzle: *the surgeon is not the boy's father.*"

Mycroft flung back his head and roared with laughter. I felt quite mortified. I had puzzled over the simple story for an hour, striving to think of unconventional possibilities: divorce, adoption, some strangely convoluted family tree—none seemed capable of explaining the facts.

"I am quite sure that there is some cheat in the wording," I said angrily.

Mycroft shook his head, still vastly amused. "No cheat, I assure you, my dear Doctor. In fact, the story is completely plausible. Given the custom of serving brewers' men a pint of ale at each pub they deliver to, I am astonished they do not end up mowing down the whole population."

He sobered somewhat at my expression. "I promise you, Doctor, the solution is quite believable. It is a simple assumption which blocks you from seeing it. Yet you will have not the slightest difficulty in accepting the answer: it is only because the assumption is so natural to you that you have not even consciously thought about it or questioned it.

"I will give you a clue: in fifty years' time, or in a forward-thinking country such as the United States, the answer may seem so obvious they will have difficulty understanding why the story is supposed to be a puzzle."

He paused for a few seconds, then said simply: "The surgeon is not the boy's father. She is—"

"His mother!"

"Quite so. And while your medical experience has led you to expect a surgeon to be invariably male, of course you can conceive of the alternative. It is always the *implicit* assumptions that trip us up."

I picked up the paper again. Two puzzles were printed each day: Mycroft had solved only the 'poser.' The 'conundrum' below was usually much harder.

"Try this one, then. An eccentric nobleman has never learned how to read a clock. When he wishes to know the time, he always asks his manservant. If he is away from home and wishes to know the time, he telephones the servant rather than reveal his ignorance to others.

"On one such occasion, he asks: 'Fanshaw, what is the time? Exactly six o'clock. Thank you,' and hangs up. But his host is in earshot and tells him: 'No, it is seven o'clock.' "

I paused for effect. "The puzzle is that *both his host and the servant were correct*. Now how on earth can that be?"

Mycroft nodded. "Again it is a question of assumptions, Doctor, and again the answer will no doubt be obvious to anyone but a dullard fifty years from now. Today it is obscure.

"You know of course that because England and Australia are on opposite sides of the world, it is always midnight in Australia when it is midday here. In fact, the Sun rises at a different time at each longitude, but it is always convenient to set the local clocks so that the Sun is overhead at midday.

"Did you know also that to the many telegraph cables which run under the Channel connecting us to the Continent, one for telephony has recently been added? No? Well, it is so.

"Now you recall that when you travel to France, the stewards come around the boat reminding passengers to set their watches forward one hour—"

"Ah, I have it. The nobleman is in France, the servant back in England!"

"Exactly so." Mycroft puffed complacently. "Has your paper no more subtle posers, Doctor? Ah well: it is rare for me to find a puzzle-setter who can truly tax me."

I saw a chink in his armor.

"There is at least one such I can think of!"

"And who is that?"

"Why, the Designer of our Universe. His little puzzle of the speed of light has you completely foxed. No doubt it is a matter of some simple assumption which, alas, you have failed to query."

"Touché, Doctor!"

Suddenly an expression of intense thoughtfulness came over Mycroft's face. His mouth half open, he drummed his fingers on the arm of his chair. Then he froze altogether, his eyes gazing into the distance.

As his silence stretched from seconds into minutes, I felt increasingly embarrassed. Would I be held responsible for reducing one of London's cleverest men to a cataleptic trance? It was with relief that I heard Sherlock Holmes's familiar tread upon the stairs. As he entered, Mycroft appeared to return to life.

"My dear Sherlock! Doctor Watson has been telling me of your proposed train experiment. I can hardly thank you enough for the idea."

"Really? It came to me that some analogous experiment must have been done long ago, it was so obvious."

"Oh, it has, it has. It was putting it into such a familiar context as a railway train that was a stroke of genius. In fact I am even more indebted to the Doctor. He has a way of making conversation which, if rather oblique to the point, can steer one's thoughts into new and profitable pathways. I really believe he has just enabled me to solve one of the most profound scientific puzzles yet encountered."

He drew the sketch overleaf.

"Here is the essence of the matter. Let us take two identical railway trains. We place them side by side, to check they are of precisely the same length, and equip each with a flashlamp at the back, and an observer with a stopwatch at the front, who can measure very exactly the time a pulse of light takes to travel the length of the train. For convenience, we will make each train one light-second long—that is, of a length which light takes exactly one second to traverse."

"Prodigiously long trains—three hundred thousand kilometers each!" I said.

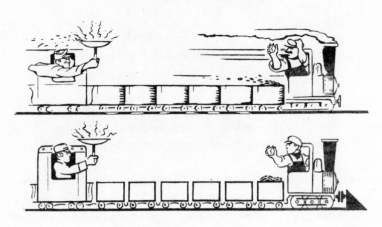

Signalling the Drivers

Mycroft smiled. "That is the advantage of thought-experiments, Doctor: one can be generous in one's use of equipment. Now we place our two trains on parallel tracks, keep one stationary, reverse the other back some distance, and then drive it forward so it whistles by the stationary train at a high speed—let us suppose, half the speed of light."

"You will certainly be the envy of any terrestrial railway engineer," Sherlock commented dryly.

Mycroft ignored his tone. "At a certain instant, the guard's van of the moving train will be precisely alongside that of the stationary one, as shown.

"At that moment, we emit a flash of light from one of the vans—"

I saw how a flaw might creep into the argument. "Ah, but is the light emitted from a stationary lamp, or a moving one?" I interjected.

"It makes no difference, Doctor. If it satisfies you, we shall place a blue lamp on the moving train, a red lamp on the stationary one, and flash them together. Irrespective of the motion, the blue and red flashes will travel together at the same speed. An observer anywhere, and moving at whatever speed, will always see the blue and red flashes arrive at the same moment. That is well established by experiment.

"Now, we ask our two observers with the stopwatches, how

long the combined light flash took to reach them. The one in the stationary cab reports, 'One second exactly.' The one in the moving cab reports, 'One second exactly.' "

I felt my mouth hang open. "But that is quite plainly impossible!" I cried.

Mycroft nodded in satisfaction. "At last you see the true horror of the paradox, Doctor. Yet that is just what is measured in practice. However, there are in fact several implicit and unproven assumptions hidden in the account I have just given. It was the riddles Doctor Watson has been reading that gave me the clue, Sherlock.

"There are at least two such assumptions here: first, that time flows at the same rate for all observers; second, that distance is the same for all observers.

"Setting aside our normal prejudices, how could the result I just described be obtained? Why, very easily, if the moving train has shrunk in length. Or alternatively, if time is passing at a slower rate on the moving train. Or perhaps some combination of those two possibilities."

"But that is absurdly contrary to all experience!" I exclaimed.

Mycroft shook his head. "Only to your own very limited and local experience, Doctor," he said. "If you spend your life all at the same spot on the Earth's surface, then you live in a world which is flat, where the time of day is the same for all inhabitants, where gravity is a force constant in both magnitude and direction, where the world beneath your feet is stable and at rest.

"As you travel farther afield, you find that none of these things is true. The world is round, the time of day depends on your position, gravity is less on a mountaintop than at sea level, the world both turns on its axis and moves around the Sun. The fuller picture unfolds only as you explore, and learn to measure your surroundings in more subtle ways than your ordinary senses permit.

"This wider perspective is but a few centuries old. And as yet our measuring instruments are crude, and our travels limited. How many 'universal constants' have we yet to discover are really variables, as we travel farther and faster?

"The contraction in space and time I am positing would be

quite tiny at speeds men can as yet attain, so unnoticeable to our senses. But if I may continue my imaginary experiment, let us see if we can be more specific.

"We have hypothesized that either time goes more slowly aboard the moving train, or its length has contracted, or both. Let us see if we can determine which is the case.

"By the way, from the point of view of an observer on the moving train, we must suppose that it is the stationary train which has shrunk, and the stationary observer whose clock is running more slowly. In the absence of an ether, there is no such thing as absolute rest, so the observer on either train has an equally valid perspective. That is a quite vital point! Let us call it the Principle of Relativity.

"Note that the width of each train remains unaffected. For if the moving train had shrunk laterally, it would fit inside the stationary train, as in a tunnel. And from the moving observer's point of view, the stationary train would fit inside his. That would plainly be a contradiction."

He took a fresh sheet of paper and sketched the picture opposite.

"Suppose that a mirror hangs on one wall of a moving railway carriage, and that opposite it is a window," he began.

"I believe there is such an arrangement on the Royal Train," I said. Some pictures of the train had recently been published; I recalled that a famous interior designer had been consulted on the layout.

Mycroft ignored me. "Imagine that we stand beside the track with a flashgun. We wish to perform the following prank: we will send a pulse of light in through the window at an angle that will, despite the motion of the train, cause it to hit and rebound from the mirror, and pass out again through the same window it entered, so it can be detected at a point farther down the track."

"Why, we used to play a similar game in my youth!" I exclaimed. "You tossed an indiarubber ball at an angle through the window of the school train as it began to pull out after you disembarked, and tried to angle it so it would bounce out again. Of course, if you missed, you forfeited the ball, unless some tolerant passenger should take pity on you and—"

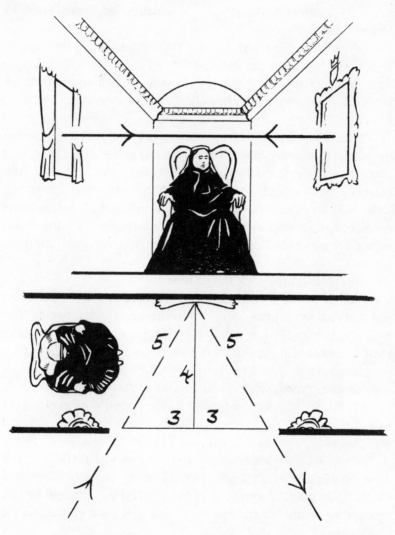

A Light Ray's Path from Different Viewpoints

"Let us add some numbers to the diagram," continued Mycroft.

"I fear mathematics is beyond me," I said rising, "so I am sure you will excuse me. I have just remembered—"

"I have promised Watson that, whatever else, I would not

permit his brain to be taxed with mathematics," explained Sherlock Holmes, smiling.

Mycroft fixed me with his eye. "Was Pythagoras's theorem on right-angled triangles beyond you?"

"As it so happens, that is just about the only theorem from my school mathematics which I can both remember and understand," I said.

"Well, then, it is the only mathematics I shall use. Suppose that the train is four meters wide. We will fire our pulse of light at such an angle that its passage to the far wall is five meters long. By the time it bounces back out of the window, it will have traveled ten meters.

"Now let us suppose that the train is traveling at six-tenths the speed of light. It will have traveled forward six meters during all this time, three meters before the light hits the mirror and three during its return."

He wrote numbers on the picture.

"Ah, it makes a three-four-five triangle," I said. "And three squared plus four squared is nine plus sixteen, which is twenty-five, which is five squared, which is correct for a right-angled triangle. How fortunate that the numbers should have turned out so neat."

"Remarkably lucky for you, Mycroft," said Sherlock Holmes with a smile. "Now I suspect you are about to tell us how this prank looks from Her Majesty's viewpoint, as the young Watson tosses his indiarubber ball."

"I should never have dreamt—" I said.

"Quite so, Sherlock," said Mycroft. "She sees a light beam enter, cross the carriage at right angles, bounce off the mirror head-on, and return to its point of entrance. It has crossed the carriage by the shortest route and bounced straight back again, a total distance of eight meters."

"So in the time we have seen it go ten meters, the Queen has seen it go eight meters," said Sherlock frowning thoughtfully. "And since the speed of light appears the same to all observers, we can deduce that traveling at six-tenths the speed of light, time flows at four-fifths its normal rate!"

"Well done, Sherlock. For those not afraid of Pythagoras, I

could write down a more general formula. Are the scribblings on that board important?"

He rose and strode to a small blackboard on the side wall, which Sherlock Holmes most often used to leave notes for myself and Mrs. Hudson.

Sherlock Holmes hesitated. "Let me see—the Palmerston poisoning—that case is no longer pertinent, yes, go ahead, Mycroft."

Mycroft scrubbed the board clear, and wrote the first line pictured on the board as shown on page 149.

"That must be the formula for calculating the rate of flow of time, at any given speed," he said.

He pondered for a moment. "Given that the speed of light seems constant to all observers, it must be the case that distance contracts in the same ratio." He wrote the second equation on the blackboard.

I struggled to make sense of this. "You mean that if I could observe such a train, standing stationary on the platform as it passed by, it would appear both squashed in length, and with those aboard it moving with dreamlike slowness?"

Mycroft nodded. "Indeed, Doctor; and the effect is present even for more ordinary trains, but to an extent too subtle for our unaided senses to perceive it. In fact you could—"

At that moment there came a peremptory knock on the door. Sherlock Holmes smiled.

"I fancy I know who that will be," he said. "When I arrived at your club, Mycroft, I was told that the good Professors Challenger and Summerlee were expected there shortly. I guessed you might have gone to ground in my rooms, and I took the liberty of leaving a note inviting them here. I really considered that if you could not solve their problem, you owed them the courtesy of telling them so in person.

"And while you have not yet resolved the wave-or-particle dispute, you have certainly demonstrated that light behaves neither like any ordinary wave nor like any ordinary particle. I leave it to your discretion whether you should tell the Professors that they are both wrong, or both right. The former might be more truthful, the latter more pragmatic, from the point of view of avoiding damage to your eardrums and my furniture.

"Watson and I would of course be delighted to assist you. Unfortunately, we have a vital luncheon engagement—not a word, Watson!—for which we are already late. Ah, welcome, Professor, Professor. You have met my brother Mycroft? I will ask Mrs. Hudson to send up tea and biscuits. And a very good day to you."

"HAS YOUR BROTHER not gone a little insane?" I asked, as Holmes and I walked toward Simpson's in The Strand. "Really, to start distorting space and time on a grand scale, to explain what is in reality a tiny discrepancy in esoteric scientific measurements!"

Sherlock Holmes smiled. "Ah, my good Watson, you like your evidence direct and unmistakable. To prove to Watson that an elephant is about, we require not merely spoor and tracks, but a large gray object with trunk and floppy ears to be standing in plain sight, or better yet on his foot. And sometimes, it may indeed be harder to deceive you than more subtle folk.

"I think we will have such plain evidence in due course. In a century's time, the dreams of Verne and Wells may come true, and fantastic man-made projectiles fly through the heavens. If they carry accurate clocks, it will no doubt then be possible to measure the distortion of space and time quite plainly and beyond dispute.

"In the meantime, Watson, do you believe Australia exists? You do? And yet you have never been there! And in the North Magnetic Pole? And in the reverse side of the Moon? There is hope for you yet. But pay attention to that hackney carriage behind you, or the evidence for its existence will become not merely apparent, but painful."

6

Three Cases of
Relative Jealousy

With a respectful knock, Mrs. Hudson's daughter Angela brought in the stack of newspapers which constituted our daily order and deposited them on that part of the table not covered by breakfast things. She withdrew with a curtsy.

Sherlock Holmes leafed through the papers with one hand, as with the other he continued to crunch toast and marmalade.

"Well, it is a tawdry crop we have here, Watson. Normally at this time of year the chief editor returns to his desk, refreshed by his vacation, and the silly season of the summer holiday period, when the temporarily promoted office junior gets to pick the lead story, is accordingly at an end. I would really have expected some more satisfying revelations.

" 'Unseasonal heat brings early harvest in Devon'—'Whale washed ashore at Brighton'—'Prince Albert joins Queen at Balmoral'—bah, what a bunch of trivia. Ah, but look, what have we here?"

He held up the front page of the *Times*. The normally immaculate cover showed signs of having been reworked in frantic haste.

Several columns of the personal advertisements which normally come below the masthead had evidently been 'pulled,' and in the space created was a slanted block of text beneath an unevenly spaced headline.

"This shows real panic in the newsroom, Watson. A duty editor's nightmare—a story too large and important for the stop press column, which comes in just as the type has been laid to bed and the print run is starting. Nothing for it but to halt the presses and set type in by hand, there on the print-room floor. Ah, I can read the signs of the emergency! Several misspellings of just the kind which occur when a journalist tries to compose directly in metal type, which is of course a mirror image of the usual letter shapes. What is the breaking story which has caused this staid newspaper to run in circles so?"

He read the story, and his face became increasingly grave. He passed the paper across to me.

"I spoke too lightly, Watson! This is no one-day wonder, as most news headlines are. This could spell real trouble in Europe. Even at this moment brother Mycroft will be early at his desk in the Foreign Office, soothing agitated ministers. And before long we will be receiving a summons from him, or I miss my guess. See what you make of it, Watson."

I read the story aloud, correcting some minor errors of grammar and punctuation as I did so.

"Central European Succession Crisis. Ullman the Second, ruler of Crolgaria for thirty years, died unexpectedly in a riding accident yesterday while hunting. His horse threw him, breaking his neck and killing him instantly. The reason for the horse's behavior is not known, but foul play was not immediately suspected.

"The Crown Prince and his wife were at the time returning from their summer vacation by the Black Sea aboard the Royal Train. Senior courtiers went to await the arrival of the train at the capital. But while it was still an hour's journey away, a double explosion occurred aboard. Both the Prince and his wife were killed instantly. They were at opposite ends of the train at the time, so an accident is hardly possible.

"It is unclear who now has the best claim to the vacant throne.

The Empire's rules of succession are normally well defined, but in the present instance, it is necessary to know whether the Prince or his wife died first. If the Prince predeceased his wife, her cousin inherits; if his wife died first, the Prince's own younger brother ascends the throne. Until the uncertainty is cleared up, considerable tension within the region must be anticipated."

I passed the paper back to Holmes.

"I can quite imagine this causing some flap at the Foreign Office," I said. "But where might you come into it?"

"The best way to defuse the crisis will be to find, promptly and with certainty, the assassin or assassins who killed the Prince and his wife," said Holmes. "And of course to establish whether the King's own death was indeed an accident. The Crolgarian police are ill equipped for an investigation of this kind. They might approach Scotland Yard for assistance, but that would be diplomatically embarrassing. Far better to send an unofficial person, who can reside inconspicuously at the Embassy, coming and going unremarked. I do not even intend to await Mycroft's summons, Watson: I should go immediately to the Foreign Office. Bid Mrs. Hudson to have a hansom flagged down and waiting for us. And now we must both change, for dress standards at the Foreign Office are high. My dressing gown will hardly do, and although you are dressed, substitution of top hat and tails will nonetheless improve the impression we make."

The cab rattled south down Baker Street, past Marble Arch, and into the grandeur of Park Lane, with its superb view over the newly landscaped Serpentine. My companion remained oblivious to the sights, staring into space and frowning. As we rolled down Constitution Hill past Buckingham Palace I saw the flag flying at half mast, in deference to the death of a fellow European monarch. When we turned into the less impressive surroundings of Birdcage Walk, by Wellington Barracks, Sherlock Holmes stirred and called to our cabbie to halt.

"See that newspaper seller, Watson? He has just had the latest edition dropped off. And we are hard by Fleet Street: the ink on the paper is still damp. Mycroft will certainly have the very latest information at his fingertips: let us show him we have not been slacking."

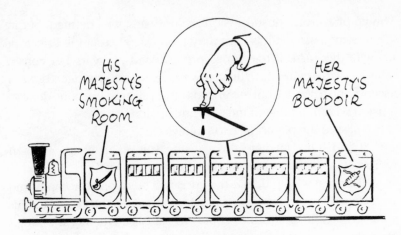

HIS MAJESTY'S SMOKING ROOM

HER MAJESTY'S BOUDOIR

The Royal Train

It was difficult to read the newsprint with the jouncing of the cab, but I could see an illustrated diagram of the Royal Train blazoned across the front page.

"There is certainly some more detail here," said Holmes. "It is confirmed that the Prince was in the front of the seven coaches which made up the train; the Princess was in the rearmost. The central coach contained the servants' station. It appears two bombs had been wired up well in advance to a cable which was used to signal simultaneously from the servant's coach to the front and rear carriages, ringing a bell in each. The assassin had removed the connections to the bells and wired the terminals to electric detonators, embedded within bales of gun-cotton.

"The problem of determining the order of death could hardly be more insuperable, Watson, if it had been deliberately contrived so! As the electric signal was sent from the central coach, even the tiny delay for the spark to travel along the wires would have been identical in each case. The gun-cotton detonated instantly. The Prince and his wife must have died at the very same moment, to within a fractional thousandth of a second. United, in a manner of speaking, in death as they scarcely were in life."

"I had wondered about that, Holmes. It seems an odd arrangement, to have had in effect two separate apartments on the train, with the only connection between them through the servants' carriage. Hardly conducive to discreet conjugal visits."

Holmes sighed. "It was an open secret that the marriage had become a complete sham, Watson. I happen to be aware of the background. It is a story which has occurred many times before, and will no doubt happen again, as long as monarchies persist.

"The Crown Prince had reached early middle age without marrying. He had had a series of friendships with women, as might be expected, but none had been approved by the Court as suitable royal consorts. Some relationships were no doubt mere dalliances, but one woman he appears to have had deeper feelings for. Marriage to her was deemed out of the question, and the Prince became bitter.

"Eventually senior courtiers decided that a suitable match for him must be found. The prime requirement of importance to them was a negative one: there must be no question of the woman having any past friendships with men on which a less than innocent construction might be placed.

"The easiest way to ensure this was to choose a very young woman, still in her teens. The introduction was arranged. The girl was attractive, and in due course a proposal of marriage followed.

"The couple were certainly ill matched, but the subsequent story depends on whether you believe the Prince's friends, or those of his wife. His camp claims that the woman was always somewhat unbalanced. Unable to provide her husband with mature companionship, she became eaten up by jealousy, and in the end turned to assassinating her husband's character with vindictive spite. Her friends say rather that the Prince was cold and unsupportive to her from the start, and soon turned to another woman without any real attempt to treat her with the respect a royal wife deserved."

"And which do you believe, Holmes?"

"I believe that one should not draw conclusions in the absence of evidence. It is in any case very difficult to assign blame for the failure of a marriage unequivocally, even if you know the parties well.

"But at all events, you understand why the couple traveled in such curious style. They were required to remain together in the public eye, but in reality loathed each other cordially. Hence the separate households, maintained even for an overnight train journey."

"I had never heard anything of this breach," I said in some surprise.

"Fortunately for the dignity of all those in public life, our newspapers observe the convention—even the more sensational ones—that they do not comment directly upon such things. I should hate to live in a world where such details were bandied about too freely. Why, success and fame in such a world would become not a reward but a positive punishment."

As he spoke, the hansom turned into the drive before the Foreign Office. We were shown into a very grand reception hall, then a more restrained but still elegantly decorated side chamber, and finally into a quite unassuming office whose only luxuries were comfortable armchairs, and a colossal desk bowed under its weight of papers.

Mycroft waved us somewhat brusquely to seats, without offering us drinks or cigars. "I am always glad to see you both, Sherlock. Alas, this is a very busy morning. To what do I owe the pleasure of this visit?"

Sherlock Holmes raised his eyebrows. "I had been anticipating a summons from you, Mycroft," he said.

"Ah, you are referring to that business in Crolgaria. I understand. Fortunately, there is no need of your assistance. I was able to solve the problem within a few seconds."

Sherlock Holmes looked disbelieving. "Really, Mycroft, I have the greatest respect for your powers. But in the tangled politics of the Balkans, there are a dozen and one factions who might have wished to assassinate any one of the three deceased. You cannot possibly be certain of the details from your desk here in London."

Mycroft in turn looked surprised. "You mean, identified the murderers? I have no idea who they were, nor frankly could I care less. I was referring to the problem of the royal succession: the question of whether it was the Prince or the Princess who died first."

"You must have better information than I, Mycroft. The account

I have here, fresh off the press, implies that the deaths were truly simultaneous."

Mycroft lifted an identical paper from his desk. "I am working from the same information as you. It is quite clear who died first."

My friend was unable to hide a fleeting expression of bafflement.

"Really, Sherlock, you should take more mental exercise. The solution comes from that very same branch of physics I was expounding when we last met." Mycroft puffed out his chest.

"The Principle of Relativity—that all frames of reference are equivalent, there being no such thing as absolute rest—implies very obviously that you can never speak meaningfully of two objects being in the same place, except if they are there at the same time."

I felt I should take Sherlock's side of this fraternal argument. "What nonsense," I said hotly. "Recently, I stood before the stone which commemorates the murder of Thomas à Becket in Canterbury Cathedral. That occurred some seven hundred years ago, yet I stood upon that same spot where he died. I swear I could feel the hair prickling upon my neck, as if from some ghostly presence."

Mycroft smiled. "Other observers might not consider it the same spot at all," he said. "Let me show you clearly what I mean."

He drew a sheet of paper toward him: it appeared to be a telegram bearing a royal crest. He turned it over and sketched upon the back (see overleaf).

"Let us suppose two planets, hurtling through space, pass within a short distance of one another. The spot ten thousand kilometers above the North Pole of the one—we shall label it Earth—coincides with that ten thousand kilometers above the South Pole of the other—we shall label that one Mars. Of course, the real planets never approach one another so closely; this is for illustration only.

"Now a year later, we ask an inhabitant of Earth to indicate that same spot. He points to a spot ten thousand kilometers above his North Pole; Mars is now millions of kilometers away. We ask an inhabitant of Mars to point to the same spot as before. He indicates a position ten thousand kilometers above his South Pole, millions of

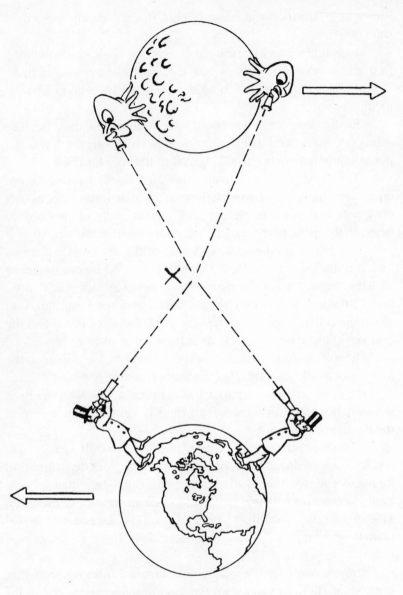

Two Planets in Passage

Signalling Both Ends of the Train

kilometers from Earth. The 'same place' is quite different from each of their perspectives."

"It is really rather a trivial fact," I said.

"Quite so! But the Principle of Relativity implies that just as we cannot speak unambiguously of 'the same place,' so also we cannot universally use the phrase 'at the same time.'

"Consider the servant at the midpoint of the train who, knowingly or unknowingly, pressed the key that blew his master and mistress to oblivion.

"Suppose he were to emit a flash of light from a torch. It travels at equal speed forward and rearward, and reaches the Prince forward and the Princess aft at the same instant, from his point of view.

"Now consider a man standing beside the railway track, who witnesses the flash emitted just as the servant passes him. From his point of view, the flash must catch up with the front coach rushing away from it, but the rearward coach is coming toward it. The flash will definitely reach the rearward coach before the foremost."

I nodded: this seemed obvious. Sherlock Holmes was evidently in deep thought.

"The situation would be unchanged, of course, if the trackside observer rather than the servant had emitted the light," Mycroft continued. "Two events which are simultaneous to the servant appear in definite sequence to the trackside observer. Now, since the forward and rearward coaches exploded simultaneously from the point of view of an observer on the train, from our trackside viewpoint the rearward coach detonated first.

"Of course, the sequence depends on the frame of reference of the observer. For example, imagine a man is traveling aboard a faster train overtaking the first on a parallel track. From his point of view, the explosion on the forward coach actually occurs before the rearward one."

Mycroft smiled tigerishly. "However we, my dear Sherlock—and any courthouse which is adjudicating upon the matter—are at rest with respect to the Earth's surface. From our particular viewpoint, the Princess died before the Prince. Unambiguously so, albeit by a rather small fraction of a second."

I felt distinctly baffled, but Holmes was nodding thoughtfully. He seemed to be calculating: his lips moved silently.

"But by such a minuscule amount, for a train moving at terrestrial speeds!" he said. "Let me see, the real interval would be some one ten-millionth-millionth of a second. An appropriately unlucky ten to the power of minus thirteen, in scientific parlance."

Mycroft folded his arms benignly.

"Did I mention that the Princess's cousin is really quite an evil man? He has even killed a servant, angry when the man failed him in some trivial task, and gone scot-free thanks to his royal protection. Some think him insane. He is quite unfit to rule a country. Whereas the Prince's brother, albeit no genius, takes his royal duties seriously, has a conscience, and is really no more likely to make a mess of things than any randomly chosen mortal.

"I have been asked to adjudicate the matter. The important thing is not the amount of time involved, but the fact that I can lay hand on heart and say that, going only and precisely by the information I have been given, the Princess would have died first."

He regarded his brother smugly. Sherlock Holmes nodded, but I felt it was time I should weigh in on his side, and perhaps dent Mycroft's complacency a little.

"Well, it is perhaps both a trivial length of time, and a rather trivial matter, to have resolved," I said. "The succession issue seems rather a storm in a teacup to me. Here in London, the royal line of a far-off Balkan country can hardly loom so important. The assassination of a king here or an Archduke there is not likely to bring about the downfall of world order, after all!"

To my amazement, Mycroft turned quite pale. He muttered something beneath his breath about the blessings of the blind. Then he appeared to recollect himself.

"Come, I am not being a very good host today. I have quite forgotten to offer you refreshment." He extended a podgy finger to a bell-button on his desk, then looked at his watch and paused.

"Come to think of it, Sherlock, I have a luncheon engagement you might both enjoy. Professors Challenger and Summerlee seem to have developed some small regard for my abilities, and Summerlee has requested a meeting. It appears it is not a matter of science that he wishes to discuss, but some problem of one of his students with complex legal implications. It could well fall more in your line of territory than mine. You are free? And you, Doctor? Capital! If we walk at a comfortable pace, we could set off immediately."

IN FACT, ALTHOUGH we made our way through St. James's Park in a most leisurely manner—physical exercise was hardly Mycroft's forte—we arrived at the restaurant in Queen's Walk somewhat early. We were examining our menus as Summerlee hurried up. He pulled up a chair and sat down, ignoring a waiter who attempted to take his coat.

"Good morning, Mr. Holmes." It was to Mycroft he spoke, ignoring Sherlock and myself.

"I am glad you were able to come. You know that as well as my teaching duties, I take my pastoral obligations toward my students most seriously—in contrast to some of my colleagues, I may add."

He spoke with some venom, and I could not help suspecting that it was Professor Challenger to whom he alluded.

"The young man in question is one Alfred Smith. He is studying physics under me. He has a twin brother named Arthur, who is also at the university here. However, he is studying music, and in fact I was unaware of his existence until recently.

"Although the brothers are identical twins, they have quite different temperaments. Alfred is much the more adventurous and assertive of the two. In fact, he has had some difficulty concentrating on his studies, amid all the distractions of youth, and last year with my approval he requested a sabbatical year. I understood he took it traveling around the world by cruise liner. This may have put some strain on his finances, for their parents died some years ago, and the family was not wealthy, although distantly related to Lord Uxbridge.

"At all events, the break seemed to do him good, for he has been much more attentive in some of his recent tutorials with me, although with odd lapses." He frowned momentarily. "No doubt the lapses are caused by brooding upon the awkward situation which has arisen.

"A few weeks ago—just after Alfred arrived back in London, in fact—Lord Uxbridge unexpectedly died. It appears he was killed by burglars, in a botched robbery at his mansion."

"Quite a coincidence of timing," remarked Sherlock.

Summerlee hesitated. "Well, in fact I know the police had their suspicions of Alfred—apparently he has some past conviction for assault. He has a volatile temperament, in marked contrast to his brother's gentleness. But he certainly had nothing to do with this crime, for I myself was able to provide his alibi. I was actually tutoring him at the moment the event occurred, many miles away. I fear your calling has made you oversuspicious, sir.

"It transpired that Lord Uxbridge had made a most peculiar will. Having no direct descendants, he was anxious that some one of his more distant young relatives should inherit the estate.

"I say some one, because he had a horror of the estate being divided. He felt that to have any chance of preserving the family tradition, a single individual must inherit. Yet he appears to have had no particular favorite among his great-nephews and nieces, none of whom he knew well. Accordingly the will specifies that the eldest such relative should inherit. But twins ran in the family, and the will specified most explicitly that the estate could not be divided in such a case. It must be determined which twin was older, by however short a time, and that twin only should receive all."

"That should resolve the matter," I said confidently. "I have assisted in the births of several cases of twins. They come one after the other, the births sometimes separated even by as much as hours, so the tie-breaking provision appears foolproof."

"Unfortunately, that does not apply in the case of Alfred and Arthur, for an extraordinary reason—" Summerlee broke off, looking through the nearby window. "Ah, here they come now. At least they are walking side by side, and seem to be talking amicably enough."

He spoke more quickly. "They must give you the rest of the story themselves. As you can imagine, the bequest has caused a deep rift between them. Each has a rather curious argument as to why he, and he alone, should inherit.

"My fear is that they will end up taking the matter to court. The business will become fodder for the popular newspapers, to the lasting shame of a noble family, and whoever wins, most of the estate will end up going to pay lawyers' fees.

"So I am hoping, sirs, that you can decide between their claims, and avoid such a disastrous outcome. Now I must leave you myself: another urgent matter has presented itself."

He rose, brushing away a waiter's attempts to offer him a menu. "Alfred, Arthur, may I introduce Mr. Mycroft Holmes, Mr. Sherlock Holmes, Doctor Watson. I very much hope these wise men will help you resolve your dispute amicably."

The young men were seated, and we made small talk as the entrées were served.

"Curious that Alfred is not distinguishable by being the more tanned, if he has just sailed around the world while his brother remained here in cloudy Blighty," I pointed out to Sherlock in an undertone. For while no adult twins are truly identical, the brothers were certainly among the most similar I had encountered, both lean and narrow-featured, with straight dark hair cut short.

"Have you not heard the slang term *posh*, Watson? It stands for 'Port Out, Starboard Home,' stamped upon the tickets of well-off travelers to India. On the outward voyage, the port side of the ship is in the shade; on the return voyage, the starboard. These are therefore the more popular cabins in tropical climes, and those fortunate enough to occupy them can avoid the Sun's rays entirely if they so wish."

He leaned forward, and spoke louder. "I understand that your great-uncle's provision that the firstborn of twins should inherit does not resolve the matter in your case. How can this be?"

Both men started to speak, but Alfred overrode his brother.

"We were technically that rarity, Siamese twins. In our case, fortunately for us, we were joined only by a flap of skin, which was easily severed not long after birth.

"But expecting a difficult delivery, the hospital performed a Cesarean section. Since we were joined in the womb, we must truly be described as having been born exactly together: as a single item, in effect."

"Then there is absolutely no way to decide which is the elder," I said in my most fatherly tone. "Clearly you must come to some amicable arrangement, or seek to have the terms of this curious will amended at appeal."

"It does not follow at all, Doctor!" cried Alfred hotly. "We were born together, admittedly. But it is not necessarily the case that we have each aged the same amount since then."

It seemed to me he was raving, but Mycroft nodded to him to continue.

"You see, I recently took a trip around the world. Six months aboard a liner, not counting various stop-offs. I traveled in the eastward direction: to Cape Town, thence to India by way of Madagascar, then to Australia, and on to Panama.

"I made my way rather painfully across the isthmus—a canal there would be a grand benefit to the world! And from the Atlantic side, took passage on a vessel to London."

"Ah, I see!" I gasped suddenly.

Sherlock Holmes smiled. "Doctor Watson here is a fan of the new brand of scientific romance," he said. "He has recently graduated from Verne to Wells, but he has undoubtedly read *Around the World in Eighty Days*, that most famous best-seller.

"Your point is, no doubt, that although you took some two hundred days to circle the Earth, you ended up having done so once more often than us stay-at-homes.

"The world turned two hundred times on its axis during your travels, and most upon its surface experienced two hundred dawns, and two hundred dusks.

"But as you went from west to east, to different time zones, your days were each a little shorter than the usual value. You ended having circled the Earth's center two hundred and one times, the additional turn by your own motion, and experienced one more dawn and one more dusk than your brother."

"Exactly so," cried Alfred, "And so I clearly am now the elder, and entitled to claim my late great-uncle's estate."

Sherlock Holmes shook his head firmly. "I am unconvinced by that reasoning. The number of seconds that have passed since your births remains identical. You could avoid dawns and dusks altogether, by residing down a coal mine, or go to the Arctic where the days and nights are each six months long. But the true duration of your age, as measured say by the number of heartbeats since birth, is not affected at all."

Alfred was about to argue further, but Sherlock Holmes held up his hand. "I think we should also hear your brother's point of view," he said.

Arthur spoke more softly, yet with determination.

"I am only a music student," he said, "but I have been following the latest discoveries about the nature of space and time with interest. Professor Summerlee was so impressed by your recent deductions," he bowed his head to Mycroft, "that he recently departed from the text of a planned lecture to explain the subject. It excited much discussion among the science students, including of course my brother, and I came to hear of the matter in some detail.

"While the mathematics is beyond me, it is quite clear qualitatively that time passes more slowly for a moving object than it does for a stationary one.

"For the last several months, I have been stationary—at least with respect to the Earth's surface—while my brother has been in motion for most of the time, at a speed of some ten knots, or near enough five meters per second.

"That may be a rather slow speed compared with that of light. But nonetheless, he has been in motion, and I at rest, and it follows that he must have aged slightly less. I am therefore now the elder, and the inheritance rightfully mine."

Alfred shook his head scornfully. "You can see my brother is at heart a musician rather than a physicist," he said in a pitying tone. "All motion is of course relative. From his point of view, I have been in motion. But from my equally valid viewpoint, it is him, and indeed the whole British Isles, which have been in motion while I stood still. I may have got older more slowly relative to him, but he has aged more slowly relative to me. The situation is still quite symmetrical."

Mycroft frowned thoughtfully. "That must be a nonsense," he said. "Suppose that your journey had been to some far star at a goodly fraction of the speed of light, in some fanciful contraption of Mr. Verne's imagination. You would return younger than your brother not by a fraction of a second, but by years. He would be stooped and aged, yourself still young and healthy. The reverse would not apply. One cannot, after all, have relative whiskers! There must be some hidden asymmetry in the situation."

He proceeded to sit, darkly brooding, until the main course was cleared away and the waiters brought desserts and coffee. He imbibed deeply of the latter—there is no doubt that caffeine, although a poison, is sometimes a useful stimulant to the brain—and at last his face cleared.

"Of course, what a fool I am!" he said. "Your clocks are equally valid only if you each continue to occupy an inertial reference frame. While a spaceship continues to travel starward at a fixed speed, this is true for both the traveler, and his friends behind.

"But at this point, no valid comparison can be made between their clocks, for they are separated. It is not meaningful to speak of a given moment—say, the moment of the spaceship's arrival at some star—as 'simultaneous' to both sets of observers.

"At that point, the traveler perceives time to have passed more slowly on the Earth, and Earthbound observers perceive time to have passed more slowly for the traveler.

"But for a direct comparison of clocks to be made, the traveler must return. And he cannot do this while staying in the same inertial frame. He must change direction and speed. Otherwise, he would simply continue in a straight line to the end of the Universe— speaking figuratively, of course: I doubt that such a place exists.

"If the star-traveler returns to Earth, he must have changed direction. He has occupied two quite different frames of reference. It is he who has therefore truly traveled, and so aged more slowly than his stay-at-home twin."

Sherlock nodded, and the twins also appeared impressed. I still felt unconvinced.

"If you wish to think it through, Doctor," said Mycroft impatiently, "imagine that the Earth twin is firing off flashes of light at regular intervals—say, of one second—for the whole time that his twin is away. Think of the rate at which his brother appears to see the flashes arrive, on his outward and homeward journeys.

"At the end, the reunited brothers will agree on exactly how many flashes there have been. But the traveler will have seen them spread over less total duration."

He looked at the twins thoughtfully.

"It is not Alfred's motion which was really important, however, but his position. For simplicity, let us pretend for a moment that the Earth's center is at rest in space—that is, it is occupying an inertial frame.

"Now the Poles are still the only points on the planet's surface which are truly at rest, for the Earth turns upon its axis. A point at the equator is moving at a speed of some four hundred and eighty meters per second, or roughly one thousand knots. The ten-knot speed of a liner is quite trivial by comparison. The turning motion is in a continually changing direction: it is not inertial, and so cannot be ignored."

"Rotation is absolute, linear motion relative," I said, remembering our adventure of the Planetarium and the Foucault pendulum.

"Quite so. Now Arthur here has been resident in London, at a latitude of some fifty degrees, and so moving at a mere three hundred and twenty meters per second, for six months, or fifteen million seconds. For all that time, his brother was close to the Equator, turning at one-and-a-half times that speed.

"At speeds low compared with light, the temporal retardation is proportional to the square of your speed. Let me see . . ."

He took out a pencil, and scribbled some figures upon his linen napkin, to the outrage of our attendant waiter.

"Yes, Arthur, by my reckoning you are some one ten-millionth of a second older than your brother.

"Incidentally, if we were to take into consideration the Earth's motion around the Sun, at thirty kilometers a second, it makes the calculation more complicated, but does not change the qualitative picture: Alfred has traveled farther, and so lost more time, so to speak."

Alfred leapt to his feet in rage. "An absurdity! No judge in the land will listen to such piffle," he shouted, and strode from the room.

Arthur also rose. "I must follow and see if I can calm him. He has always been impulsive. Well, sirs, it seems I may inherit after all, and if so I will make full provision for my brother, of course. But I fear it will not be without a considerable legal battle, and much ill will."

A sparkle had come to Sherlock Holmes's eye. "One moment, before you go," he said. "I fear my brother has missed one rather major point."

Mycroft considered. "I think not," he said firmly. "I am quite sure of my reasoning."

"I daresay. But you have nevertheless overlooked something." Sherlock Holmes turned to Arthur. "You seem quite knowledgeable of physics, for one who is studying music. I have known twins before. Would I be correct in supposing that Alfred has occasionally asked you to attend lectures on his behalf, signing the register and in effect impersonating him."

The young man blushed. "Alfred is tempted by the many distractions of London," he said. "Yes, I have done so on a few occasions. No harm has resulted, for I have taken notes for him quite conscientiously."

"I should even think you could get away with attending the occasional tutorial in his place," Sherlock continued remorselessly. "I cannot imagine that Professor Summerlee is the most observant of men, as pertains to the features and mannerisms of his students."

"I promise you that I have done that only once. Alfred was suffering a severe hangover and—"

"Would that by any chance have been on the day of Lord Uxbridge's death?" asked Sherlock swiftly.

"Why, yes. Now however did you guess that?"

"You will find out quite soon. Now perhaps you had better follow your brother."

The young man left us. Mycroft was blushing deeply; Sherlock smiled at us both as the waiter cleared the table.

"Do you not see, Watson? Alfred was suspected of being responsible for his great-uncle's death. His alibi turns out to have been provided, in all innocence, by his twin! An excellent alibi, too: no one would question the word of the esteemed Professor Summerlee. We must do a little conventional investigation to clinch the matter, but I have little doubt what we shall uncover."

He rose, and I made to follow him. Mycroft still sat rigid.

"My dear Sherlock," he said stiffly, "I would of course have seen the alibi business at once, but I was so preoccupied with the paradox that—"

Sherlock Holmes smiled mischievously. "There is no need to apologize, Mycroft," he said. "Science is a fascinating study, but if there is a moral here, it is that one should never let it blind one to matters of mere human concern, such as motives and morals."

I followed Sherlock out into the sunlight.

"It never does any of us any harm to be reminded that we are all but fallible humans, Watson," he said cheerfully. "In this case, one may not profit from the proceeds of a crime, so poor Alfred has neatly ruled himself out of consideration, even if he is spared the gallows.

"In fact, your own astute observation about the lack of a tan should have given Mycroft a clue, if one were needed. Notwith-standing my remarks about port and starboard, while it is *possible* to traverse the tropics without acquiring a suntan, it is quite remarkable that an active young man should do so. He must have confined himself to the shade like an invalid. The idea for his alibi must have occurred to him at least some weeks before the end of his journey. Very much a premeditated crime, Watson! Now let us proceed to its trail. Be so good as to help me attract the attention of that cabbie."

7

The Case of the
Faster Businessman

It was at the end of a sweltering August day that I trudged homeward along the Marylebone Road. The genteel Notting Hill area where so many of my patients live is separated from Baker Street by rougher parts, and I had to pick my way past lolling dogs, tongues hanging out as they panted, and festering mounds of garbage. I had made the unwise decision to perform my rounds on foot, and as No. 221B at last came into sight, the vision of soda-siphon and armchair rose enticingly before me.

The scene which I encountered in the hallway was therefore less than welcome. Mrs. Hudson stood, arms folded, blocking the stairway from a tall, expensively dressed man of late middle years and choleric features whom she had evidently been confronting. She looked at me appealingly.

"Doctor, I have just taken this gentleman's card up to Mr. Holmes, who has made it quite clear that he is very busy and has no time to see him. But the gentleman will not take no for an answer, and even tried to push past me just now. Perhaps you could explain to him?"

I drew myself up as sternly as possible, in the wilting heat. "Mr. Holmes is a very busy man, with a large consulting practice," I said coldly. "I am his partner, Doctor Watson, and if you would be so good as to explain your case to me, I will perhaps be able to arrange an appointment for you later in this week, or possibly the next. Whom have I the honor of addressing?"

The man's face turned almost white with anger, in unpleasant mottled contrast to the purple of his neck.

"Next week!" he shrieked in a strong New York Bronx accent. "By heaven, it is no wonder that you Brits are drifting into decline as we on our side of the Atlantic leap ahead. Have you no sense of the value of time? My name is Barnum Rolleman, and I am one of the richest men in America. Now you tell your master that if he does not see me immediately, he will be very sorry indeed!"

I stiffened further. "Speaking as a medical man, I would caution you not to excite yourself so: it puts an inadvisable strain on the heart. I can assure you that threats will get you nowhere, and if you wish us to consider your case at all, I suggest that you—"

Rolleman had been fairly trembling with rage, but now visibly pulled himself under control and spoke in a lower tone.

"I was not threatening you, sir. I was referring to the fact that I am a wealthy man, and I have found it always pays to get the best advice. I am being blackmailed, in a way that the police cannot help with. Rather than pay the blackmailer, I am willing to give the private detective who resolves the case for me a tidy sum—a very tidy sum indeed. I am sure Mr. Holmes would regret passing up such an opportunity."

"I know that Sherlock Holmes disapproves particularly of blackmail," I said. "I will try to persuade him to see you. But if he is willing only to give you a later appointment, or none at all, then you must accept that."

He barely nodded, and I went on upstairs to our rooms. Sherlock Holmes was sprawled across the sofa in a dressing gown, idly doodling on a sheet of paper.

"Really, Holmes," I said severely. "I hardly think you look like the busiest of men. I can see why Mrs. Hudson is a little upset at being used as a doorstop. Is it not worth your while to at least interview this wealthy client?"

He waved a languid hand. "I find business moguls rarely bring me interesting cases, Watson. Usually they are concerned about some matter of misappropriation or financial goings-on within their empires which are more a matter for an accountant than a detective.

"I have known Mycroft to investigate suchlike matters, but only on behalf of the government. He does not prostitute his talents. And I really cannot get excited over columns of figures, on behalf of men who already own more wealth than they possibly have a use for."

"You are wrong in your guess, Holmes," I said. "Rolleman is being blackmailed. Even you cannot leap to conclusions without bothering to look into the matter at all. It is your perfect right to refuse his case, but I really think you could do so in person, rather than pressing your household into service as intermediaries."

I had spoken quite firmly. Holmes looked at me and sighed. "Very well, Doctor. If I must sacrifice a few moments in the interests of subsequent peace and quiet, so be it. Show Mr. Rolleman in."

I went out onto the landing before he could change his mind, and beckoned to Rolleman, who was pacing the hallway below. He fairly sprinted up the stairs, and scarcely paused at the top although clearly short of breath. We entered to find Holmes had made no attempt to straighten up his posture. He lazily indicated a chair.

"Mr. Rolleman, I am honored," he said in tones of faint irony. "To think that a man of so many millions should visit our humble abode. Pray, how might I be able to assist you?"

Our visitor looked at him in disgust. "Well, it is easy to see why I am a man of wealth, while you despite your cleverness are still in quite ordinary circumstances," he said. "It is not cleverness that conquers the world, Mr. Holmes, but energy. Energy and above all swiftness in action, that is the secret to riches and success. Do it faster rather than better, and you win."

Sherlock Holmes raised an eyebrow. "Myself, I prefer to think that human advance depends more on laziness," he said. "Take the inventor of the wheel, for example. I would wager that he was no ambitious go-getter, but merely loathed the effort of dragging the kill back to his cave over his shoulder. 'Avoid this toil, and I could enjoy more idleness,' he thought. And so we obtained the cart.

"Then there was the man too lazy to chase swift game. 'If I had a means of throwing my spear more swiftly than the prey,' he no doubt reasoned, 'I could relax and grow fat while others run frantically about.' And so we have the bow and arrow. Then again—"

I could scarcely blame Mr. Rolleman for interrupting, as it seemed Holmes was prepared to recite this curious interpretation of the history of mankind for several thousand years' worth unless stopped forcibly.

"Your example of the arrow over the spear is a good one for me, Mr. Holmes, for my whole empire is built on just that advantage of improved speed. I saw that the paddle steamer would overtake the sailing ship, bringing my goods more swiftly than the next man's, and I made my first fortune. I consulted the best scientists and engineers who told me that the screw-drive propeller would nonetheless overtake the paddle wheel in due course, and I made my second. I used the steam railways to take my goods inland while others relied on the horse and cart, and my empire grew America-wide.

"More recently, I spotted that the telegraph's ability to send messages instantly could be vital to stealing a march on one's rivals. I persuaded the big dealer markets to do business instantly by electricity. That brought me still more money, but also a threat. And so I come to you."

Sherlock Holmes looked mildly intrigued. "Yes, you mentioned blackmail," he said. "Please continue."

"My most profitable business venture nowadays is the Superior Exchange, based in Chicago. This is a brokership which allows men at the Stock Exchange here in London to place bids for bulk commodities, in a fashion similar to an auction house.

"It is vital to the integrity of the business that all potential bidders receive information at precisely the same time, to ensure fairness. Bids are then accepted on the basis that the first received wins if there are two equal offers, as very often occurs.

"The telegraph system proved inadequate for our purposes.

To ensure all European buyers receive information at exactly the same time, we use a specially adapted version of Mr. Marconi's recently perfected apparatus. A signal is sent into the ether from Chicago to announce the bidding open, and traveling at the speed of light is received in London just one-sixtieth of a second later. Those brokers who wish to bid at that point immediately send back a response by the same means. Each bidder is assigned a different radio frequency.

"Of course, there is intense competition among the London brokerage houses to signal their bids as fast as possible. Some months ago, they stopped relying on human operators to do so. Now when they want to signal a bid on the next sign from Chicago, they have but to hold down a key, and a return pulse is sent automatically and instantaneously. Electric apparatus at the Chicago end can distinguish the first return signal received, even if they are separated by a thousandth of a second or less, and an electric bulb illuminates to show the identity of the successful bidder."

Sherlock Holmes shook his head, "Such ingenuity, that rich men may play a form of up-market poker the more swiftly!" he said.

Rolleman ignored him. "I was concerned right from the start that there might be a loophole in the system," he said. "Suppose some broker was able to anticipate the radio sign from Chicago, then he could jump the gun. For example, suppose some accomplice in America had a radio receiver near Chicago, connected to one of the transatlantic cables which carry telegrams. The accomplice could operate a telegraph key, sending a signal to London which might get there before the radio wave, it seemed to me.

"However, I was assured by the most expert American scientists that such a thing was impossible. No electrical signal, whether down a wire or through the ether, appears able to travel faster than light. So no such fraud should be possible, even in principle.

"Until a few weeks ago, my confidence in the integrity of the system was complete. Then I received the following letter, delivered in Chicago with a London postmark."

He passed it across to Sherlock Holmes, who read aloud:

Dear Sir:

I am pleased to inform you that I have discovered a simple method of transmitting messages faster than light. Being a responsible person, I wish to avoid the disruption in business confidence, especially to your brokerage, that publication of the method would bring.

However, I would expect some reasonable compensation for forgoing the income my invention might otherwise yield. I will be happy to negotiate terms. Please signal your readiness to do so by placing the message "Mandrake Is Courted" in the Personals section of the *Times*.

Yours etc.,
Mandrake.

Sherlock Holmes looked up. "Is this the note you refer to as blackmail?"

"It is."

"Well, I would hardly call it blackmail in the legal sense. Selling an invention, or even suppressing one, is quite as legal as your own business activities. Why do you not simply offer to purchase this Mandrake's device?"

Rolleman shook his head. "I have already been in contact. We have exchanged several messages. He refuses to sell, and will offer only to suppress the invention, without giving details of it beyond hints incomprehensible to me."

"Well, surely you should consider that he is bluffing, and it is time to call him out!" I said.

"I dare not do that. I understand there have been rumors of new discoveries pertaining to the speed of light and related phenomena circulating in London—"

"We are well aware of them," commented Holmes.

"And if Mandrake goes public, however ill founded his claims, it could be the ruin of my brokerage. Yet I am unwilling to pay him the sum demanded on a basis that is, as you say, most probably mere bluff.

"What I need, sir, is one of two things: either information that it is possible to send a signal faster than light, with some indication how that feat might be accomplished, or else an assurance that it is definitely impossible and will be so for all time."

Holmes raised his eyebrows. "You have been unable to find a scientist willing to provide this information?"

Rolleman smiled ferally. "I could easily have done that. But I have my own unique way of paying for such opinions.

"I like to think I am not quite a fool in business matters. I discovered many years ago that the world is full of consultants of one sort or another—lawyers, scientists, accountants—who are willing to give glib advice which turns out to be founded on sand. I found a simple solution. Nowadays I pay for advice not in the form of a fee but in the form of a wager."

He drew a checkbook and fountain pen from his pocket, and wrote. He held the check up for us to see.

"This is a note for twenty thousand pounds, made out to Mr. Sherlock Holmes."

I could not restrain a gasp.

"But I will give it to you only in return for a signed wager. You may bet either that you can tell me a method of sending a message faster than light, or that such a feat is absolutely impossible. The rider is that if you are ever proved wrong, you must not only return my twenty thousand, but as much again, as loser of the bet.

"I have consulted several scientists, and although they all think faster-than-light signaling impractical, I have found none to take my wager."

Holmes looked amused. "I might take you up, Mr. Rolleman. But I will need time to think it over. You are staying at a hotel? The Savoy? Good: we will call upon you tomorrow. Watson, be so good as to show Mr. Rolleman out."

I returned to Holmes in some alarm. "You cannot be seriously thinking of taking his wager!" I exclaimed. "Why, it would ruin you if you lost. What an iniquitous way to pay for advice."

"On the contrary, I think it is rather a fair way. Imagine how rapidly the world would be rid of dissembling lawyers and incompetent accountants if all consultancy was on such a basis! But fear not, Watson, I shall not be rash. I can call in some favors, and obtain advice on the subject from three of London's cleverest men. If you could be patient while I dress."

As I waited, I thought about the matter. Proving a negative is always problematic: I was sure it was quite beyond my abilities. But finding a way to send a message faster than light struck me as more straightforward. Could I not devise some arrangement of mirrors and lamps which would do the trick?

I greeted my friend's emergence triumphantly. "Do not trouble to don your hat and gloves, Holmes. There is no need to go out. I have solved the puzzle."

I showed him the diagram which I reproduce below. I had striven to emulate Mycroft's clear drawing style.

"This central tower represents a lighthouse, Holmes. Now a lighthouse appears to the mariner to flash at regular intervals, but in reality it emits a narrow beam which rotates as a focusing lens or mirror is turned about the central lamp. At the moment I show, the beam is emanating westward. Suppose that the lighthouse is emit-

Doctor Watson's Lighthouse

ting one flash per second, that is, the beam is rotating at a corresponding rate.

"Let us imagine the lighthouse is surrounded by a low circular wall at a distance of one kilometer. The total length of the wall is just over six kilometers, and the spot of light moves along it at six kilometers per second. Already it is going faster than the fastest shell fired from a gun.

"Now, enlarge the wall's radius to one thousand kilometers. The beam sweeps it at six thousand kilometers per second. Increase the radius further to one hundred thousand kilometers. The spot moves at six hundred thousand kilometers per second—twice the speed of light!"

Sherlock Holmes frowned.

"Your diagram is a little misleading, Watson. For one thing, you are forgetting the basic fact that light travels at finite speed. If you could see the illuminated region from a point very high above, it would actually take the form of a spiral. Rather like the jet of water from a rotating lawn sprinkler."

He drew the diagram below.

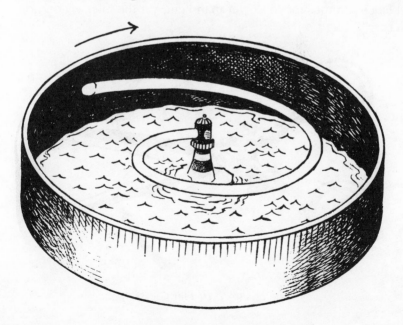

Sherlock Holmes's Lighthouse

"Nevertheless," I insisted, "the spot still moves along the wall at twice lightspeed—or indeed any speed I like, if the wall is sufficiently distant."

"Ah, but we need to use it to send a signal. How would you employ the beam to send a message from one point on the wall to another?"

"Well, I suppose a person on the wall could reflect the beam back toward the lighthouse with a mirror—" I said. Then I stopped, seeing the objection.

"Quite so, Watson: your message will always take longer, or at best just as long, as using light transmitted directly."

I refused to give up. "Notwithstanding, I have a second scheme here which does not depend upon the properties of light at all." I drew the diagram below. "This represents a pair of scissors, Holmes. A very large pair of scissors, worthy of Mycroft's thought-experiments.

"Consider the point of intersection as the blades move across. Clearly this point can move much faster than the blades themselves, depending on the narrowness of the angle between them. The blades can thus move slower than light, yet the intersection point faster."

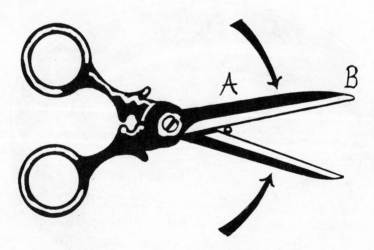

The Faster-than-Light Scissors

Holmes considered. "And how will you use the effect to send a message?"

"In the following way. I place at **A** a small hard blocking object, of the sort that ruins real pairs of scissors. The scissors close freely until **A** is reached, then stop. **B** is almost immediately aware of the stoppage: thus a signal can be sent."

"Bravely tried, Watson. But I fear there is still a problem. Real substances are not infinitely hard and rigid. If you hit one end of a ruler, for example, the other end does not move instantly but a fraction of a second later. In fact, vibrations travel through solid objects at fixed speeds, just as sound travels through air at a fixed speed. For the hardest substances known, the speed is about ten times faster than sound in air: still one hundred thousand times slower than light. The effects of any impact at **A** will therefore not be transmitted to **B** until a correspondingly later time."

Sherlock Holmes clapped me on the shoulder. "Well thought, all the same! But I think this problem is one for the experts. Let us set forth. We will call first upon Professor Summerlee."

HOLMES LED THE way briskly along the Euston Road, then into a side entrance of University College surrounded by high walls of gray stone. This was evidently the quad devoted to the science faculty. I paused before a stone arch with Greek lettering above it.

" 'Let no one enter here who is not competent in mathematics,' " I translated.

"That was supposedly the inscription over the gate of Plato's school of logic, two thousand years ago," said Holmes. "Mathematics then was a rudimentary discipline. Someone such as yourself who could perform division sums, and even knew Pythagoras's mystic formula, would not only have qualified, but quite impressed the gatekeepers."

"Nevertheless, I feel like a heretic entering a cathedral: the keystone may fall and crush me as an impostor!" I said jokingly.

At that moment there came a rattle of wheels on cobblestones, and a side door opened to permit passage of a flat barrow laden with baked pies, evidently destined for the college buttery.

"This way," cried Sherlock, and we ducked past the surprised porter and along a corridor. "I would hate to be responsible for your premature demise, Watson, and Summerlee's office is in any case located on this side."

Moments later, we knocked and were admitted to a small bare room. Three walls were covered by blackboards scrawled with incomprehensible symbols and equations. Summerlee listened impatiently to our request.

"It is definitely impossible to travel faster than light," he said briskly. "You recall the formulae for the relativistic contraction of space and time?"

He pointed to one of the blackboards, where I saw that Mycroft's formulae had been reproduced (see page 149).

"A speed faster than light implies a Beta of greater than one," he said. "That yields a factor for the contraction of space and time which would be the square root of a negative number."

"But I thought that minus one—or indeed any negative number—had no square root," I said.

"No *real* square root," said Summerlee sternly. "A negative real number multiplied by a negative real number yields a positive result, as of course does a positive multiplied by a positive. So no real number, positive or negative, squares to produce a negative result. We can define an *imaginary* quantity whose square is negative, but it is a mathematical fiction.

"Since the formulae give results which clearly cannot pertain to the real world, it follows that it is impossible for anything to occupy a faster-than-light frame of reference."

"I do not quite see—" I began.

"Possibly not, but that is the rigorous mathematical answer to your question. And now I do not wish to be rude, but I am afraid duties more important than the settling of wagers await me."

A HALF HOUR later, after a stroll across Hyde Park, we found ourselves in the more open surroundings of Imperial College, where we were directed to Professor Challenger's study. As we reached the closed oak door, the Professor's irate voice boomed out. He appeared to be lambasting some unfortunate student. At length the door opened, and a young man appeared.

"Is Professor Challenger free?" Sherlock asked.

"You *wish* to see him?" asked the young man incredulously. Then recollecting himself, he nodded and hurried away. We entered the Professor's den with some trepidation, at least on my part. Sherlock explained the wager, and the answer Summerlee had given us.

Challenger sat back in his chair and roared with laughter.

"A true mathematician's answer," he crowed when he was able to speak. "It may indeed imply that faster-than-light travel is impossible, but it hardly gives a useful picture of *why* it is impossible, or *what* would happen if you were to attempt it. So many mathematical proofs are like that. They may be rigorously valid, yet give no useful picture of the situation they describe.

"There is an apocryphal story of a mathematics student who, to broaden his scientific perspective, was assigned to a geography project. He was asked to determine by investigation whether a given place was on an island or on the mainland. It was in fact on the island.

"He shuffled about the mainland, gazing at his shoelaces. Now as it so happened, a tunnel had been dug connecting the island to the mainland, and by chance he wandered into it. He never looked up to notice he was in a tunnel, nor would he have thought it relevant if he had. Shortly, he stumbled across his destination. Since he had reached it by walking on dry land, he reported it was on the mainland.

"And in a sense he was right. You can ask a mathematician about islands only if you provide him with a rigorous definition of an island, and the same applies to any other topic. And by some definitions, install a tunnel or bridge to an island and it ceases to be one! But his answer was hardly helpful, and he had learned nothing useful about the world around him.

"So, let us take a more practical approach to your problem. Let us suppose an engineer wishes to fire a projectile faster than light. The best naval guns can fire a shell at about one kilometer per second. But if we were to mount such a gun on a vehicle already traveling very close to light's speed, could we not overtake its rays?

"Imagine your brother Mycroft's hypothetical train traveling at six-tenths the speed of light has such a gun mounted on it, pointing forward. It fires. From the point of view of an observer standing by the track, how fast does the shell travel?

A light dawned on me. "Aha!" I said. "From a trackside viewpoint, the train is compressed in its direction of travel, and moreover, time aboard it is passing at a leisurely pace, by a factor I recall: four-fifths of normal in each case. Multiplying these effects, the shell will be traveling only six hundred and forty meters per second faster than the train."

Challenger beamed at me. "Very good! But our engineer is persistent. He raises the speed of the train to just three hundred meters per second below the speed of light—one millionth part less—and fires his gun. Has he any success?"

I saw the formulae for relativistic space and time compression on a nearby blackboard, pictured opposite.

"Why, events aboard the train will seem to be happening at a snail's pace, and additionally the gun and bullet will be squashed to a hair's width!" I said.

"Quite so. The shell would barely crawl forward, and certainly not overtake light. In fact, the method you used to add speeds is valid only when one of those speeds is small compared with that of light. But it is easy to devise a more accurate formula." He wrote the bottom line upon the blackboard. "However large your initial Betas are, as long as each is less than unity, the sum will also be less than unity."

He raised a podgy finger. "You are no doubt wondering where the energy of the gunpowder involved is going. The answer is that it is indeed conveyed into the bullet. In a relativistic world, energy is not simply proportional to the square of the velocity, but increases exponentially as light speed is approached.

"To propel your shell, or any other object, to lightspeed would take an infinite amount of energy. Which, I would venture to say, imposes a fundamental limit on our engineer's endeavors."

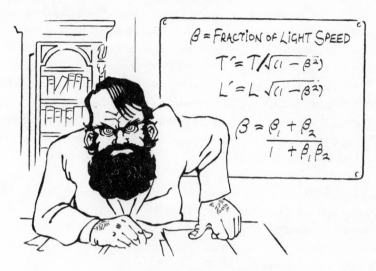

$$\beta = \text{FRACTION OF LIGHT SPEED}$$
$$T' = T/\sqrt{(1-\beta^2)}$$
$$L' = L\sqrt{(1-\beta^2)}$$
$$\beta = \frac{\beta_1 + \beta_2}{1 + \beta_1\beta_2}$$

The Relativity Equations

I thought of Mr. Wells's novel *The First Men in the Moon*, which I had just finished. I had been reading that the stars were incomparably farther than the Moon, and wondering if we could ever travel to them.

"The nearest star to our Sun is believed to be so far away that its light takes five years to reach us," I said. "You are telling me that I could never travel there in under five years, aboard however wonderful a craft."

Challenger smiled cunningly. "From your point of view, you could," he said. "As you accelerated, you would see the Universe itself appear to squash up in the direction of flight. So although you would never perceive yourself as traveling faster than light, you could indeed get there in a shorter time."

"It would be as if space itself were warping!" I said.

"Not exactly warping, my good Doctor, but squeezing itself up in a regular fashion, would be a more accurate description."

Sherlock Holmes nodded. "But for an observer back here on Earth, the round trip would still take at least ten years, which is the relevant point for our proposed wager. I can quite accept that no *physical* object can travel beyond lightspeed. But I am less sure about other manifestations. Can you tell me for certain, Professor, that no new kind of radiation or thing might be discovered, which could be transmitted faster than light?"

Challenger pondered. "No, I cannot be so definitive," he said at last reluctantly. "A wise man never assumes that the as yet unachieved is impossible, knowing the limits of his own knowledge. Who knows what may be discovered tomorrow? I could not possibly guarantee such a thing."

"I FEAR WE are on a wild goose chase, Watson," said Holmes as we left the College. He glanced at his watch. "But it is near the cocktail hour, and Mycroft will assuredly be at his club, which is nearly on our way home. Let us call in, although I am sure he will echo Challenger's opinion."

But Holmes, for once, was wrong. Mycroft looked at us with keen interest as the problem was explained.

"I can tell you quite definitively that faster-than-light signaling is impossible, by any means," he said. "You are a Wells fan, Doctor. Have you read *The Time Machine*? You enjoyed it. But did you consider it plausible?"

I almost laughed aloud, before I remembered how grave a breach of etiquette this would constitute at the Diogenes Club, even in the Strangers' Room.

"An amusing fantasy, but obviously impossible," I said. "Why, if I could travel back in time, I could go back and shoot my grandmother before she had children. So then I would not be here today, to go back and commit the deed! Or I could send a message back to myself, that I knew I had not received. No end of paradoxes would arise, if it were possible even to communicate backwards in time."

Mycroft nodded. "Very good, Doctor," he said ponderously. "Now consider that imaginary train of mine you are familiar with, one light-second long. Let us equip the driver at the front and the guard at the back with some kind of super-telegraph. Normally a signal would take at least one second to pass between the two, but we will suppose this telegraph works instantaneously.

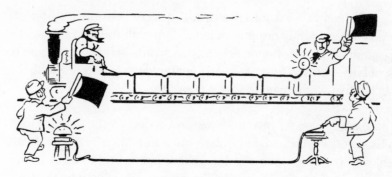

Signal Arriving Before It Is Sent

"Next, station two men beside the track, at a distance apart so that the train's engine is alongside the first just as the guard's van is alongside the second, from their point of view. These two men are also linked by a second copy of the super-telegraph.

"Now consider the following sequence. The forward trackman gives a signal to the driver as he passes. The driver transmits the signal to the guard.

"Here is the rub. You recall that if two events happen simultaneously from the driver's and guard's viewpoints, for trackside observers, the guard's event occurs earlier than the driver's. The Princess died before the Prince. The guard therefore receives his message more than one second before it is sent, from a trackside viewpoint!

"The guard signals to the rearward trackside observer as he passes. The rearward observer passes the signal back to the forward observer, who sent it in the first place."

"We had to do these Chinese-whisper signaling exercises while I was in Afghanistan," I said. "Passing a message from man to man, and seeing if it was garbled by the time it had done the round. There was a rumor that the message 'Enemy advancing, send reinforcements' came back as 'Emily's aunt's died, send three and fourpence.' "

Mycroft looked at me severely. "Well, something still more remarkable happens in this case," he said. "Do you not see? The signal returns to the forward observer more than one second before he sent it! An obvious paradox has occurred.

"In fact, it is easy to see that, if it were possible to send any signal even slightly faster than light, then with the aid of a suitably fast train—or pair of signalers, in other words—a message could be sent backwards in time. As sure a guarantee as you could wish that such a feat will remain forever impossible."

SHERLOCK THANKED HIM, and we proceeded to the Savoy Hotel. Visions of wealth passed through my mind. My friend had always enjoyed the chase over the kill, taking interesting cases in preference to lucrative ones. What difference to his style would twenty thousand pounds make?

A concierge took us up to Mr. Rolleman's suite, and knocked at the door. There was no reply.

"He is definitely in," she said. "He passed me a short time ago. He seemed unwell, almost gasping for breath."

She opened the door, and shrank back with a scream. Mr. Rolleman was lying almost at her feet, his eyes staring blindly upward from a face of ghastly purple. I knelt and examined him as Holmes moved swiftly about the room, looking for signs of intruders.

"There is nothing here for you to solve, Holmes. It is clearly a massive apoplexy. I saw the signs earlier today: I cannot be the first doctor whose advice to slow down he ignored."

"How often true is that trite old saying: 'More haste, less speed.' At least in the sufficiently long run," commented Holmes as we left. "Mr. Rolleman was in a sense right in his opinion of me: I am by his standards a dilettante. I apply my talents to detection because I enjoy it, even if the same effort in another field could have made me wealthier. I live in lodgings humble by the standards of some, because they suit me. But I pass my time comfortably, in a style that fits me like a well-worn pair of slippers, with just the right amount of excitement and adventure for seasoning. And since Rolleman's fate awaits us all in the end, Watson, what more can any of us really aspire to, than a life lived at the right speed?"

8

The Case of the Energetic Anarchist

"WELL, I SHOULD NEVER HAVE dreamt it," I exclaimed. "A significant application of mathematics to medicine, and moreover in a way that I can comprehend."

I held up my copy of the *Lancet*. "It has always been a mystery, Holmes, how a malady long present in a country, but causing just a handful of cases each year, can turn almost overnight into a pandemic: a plague running through the land like wildfire, until within a few months all citizens must be either dead or immune.

"Many brave doctors have traveled to such countries in a vain attempt to understand the phenomenon, sometimes themselves becoming infected and dying. But now a consultant resident here in London has come up with a key insight.

"Supposing, he says, that humans are only occasionally infected with a pathogen residing in some other species: for example, there might be ten such cases each year in a given country. If the disease also has a human infectivity, an ability to pass from person to person, those ten are multiplied by a certain factor. For example, if the infectivity is point five, in other words, each infected person has a one-half probability of infecting one other person before he or she recovers or dies, then each root case has a one-half chance of triggering a second, plus a one-quarter chance of triggering a third, and so on in diminishing sequence. The multiplying factor is two, and there will be a total of twenty cases per annum.

"Even if the infectivity is as high as point nine nine, any pocket of infection will die out quite fast. A total of one hundred will become ill for every initial case. One thousand cases per annum would still not be high, for a large tropical country.

"But now suppose the pathogen becomes the tiniest fraction more virulent: say the infectivity becomes one point zero one. Now instead of tending to die out, each pocket of infection grows remorselessly. In due course each case leads to two others, those two to four, and so on. Behold: plague on a biblical scale!"

I looked severely at my colleague. "Now here is an application of the sciences that really matters, Holmes. Unlike, if I may say so, the dabblings of your brother in the theology of physics. Timings in error by millionths of a second! Anomalies of distance that become apparent only at speeds far higher than man can ever hope to achieve! The new physics is but a toy, Holmes: the discovery of relativity will make not a ha'penny's worth of difference to any practical thing."

Holmes looked at me. We were recovering from our exertions of the previous week, browsing through the weekend's delivery of periodicals: for myself, medical journals; for Holmes, the penny dreadful Sunday tabloids.

"It is a good insight, Watson. I wonder if a similar process could explain the spread of rumors." He tapped the front cover of the paper he was holding. The enormous headline ANARCHISTS POISED TO STRIKE had the subheading 'Could Your Best Friend Belong to a Secret Society?' Below that was a line drawing of a man dressed in a top hat, striking a match above a spherical cartoonist's bomb, with a dreadful leer upon his face.

"For example, at the moment we are suffering a plague of anarchists. Not the real thing, of course, but rather a pandemic of stories about anarchists and conspiracies and such. Each story builds on the last, exaggerating it by just a little, and soon even people of respectable intelligence are half believing it all. I had a solemn letter from Scotland Yard yesterday, asking me to stand ready to assist in the crisis. And who are the real anarchists, if any? Probably a group of students who amuse themselves by writing anonymous letters to the papers. I tell you, Watson—"

At that moment, he was interrupted by a diffident knock. We looked up to see a telegraph boy standing in the open door.

"Telegram for Mr. Holmes, from Scotland Yard," the lad said with great importance.

Holmes took the telegram. "And on a Sunday!" he said as he slit the envelope. "The world must really be about to end, Watson." He read the enclosure, then tossed back his head and laughed. He handed a sixpenny to the telegraph boy. "Thank you, lad: no, no reply."

As the boy clumped down the stairs, Holmes passed the telegram to me.

" 'LATEST TO TIMES FROM ANONYMOUS STOP THREAT TO EXPLODE BOMB OF POWER EQUIVALENT TO ONE HUNDRED THOUSAND TONS OF GUN COTTON IN CENTRAL LONDON TUESDAY STOP PLEASE ADVISE STOP ARNDALE,' " I read.

"Arndale is the duty man at the Yard. Even Lestrade would not fall for that one!" said Holmes. "And the letter writer is himself naive. He evidently expects Monday's *Times* to print it, which of course it will not, and start a panic of refugees fleeing before Tuesday's deadline."

"Well, I hope you are sure!" I said.

"Consider, Watson. Suppose you wanted to smuggle the maximum quantity of high explosive into London; how would you do it?"

I thought about the problem. "I remember the story of the Trojan Horse, Holmes. I think I would take a train into a siding somewhere and load it with high explosive. Then I would arrange to substitute it for a legitimate train to a London terminus, perhaps an early-morning milk train. No one opens a milk churn to see if it contains milk or explosive."

Holmes nodded. "Quite ingenious, Watson. But your plan

would be vulnerable to interception: a train is too much at the mercy of signalmen and line blockages. And you would get a few hundred tons of explosive in at the very most.

"I would choose rather a barge convoy. The tugs which tow massive blocks of roped barges up and down the Thames daily are so familiar we hardly notice them. You might get a thousand tons or so aboard. And you could get it to a point more central than any railway terminus, right alongside the Houses of Parliament if you wished. I might take such a plan seriously. But a hundred thousand tons is absurd. I will not disrupt my Sunday to amuse some half-wit."

I thought of a possibility. "Might there not be some explosive which is, weight for weight, much more powerful than gun-cotton?" I asked.

"I do not think so, Watson. Gun-cotton is just about the best such substance known." He considered. "But to set your mind at rest, and convince the Yard I am earning my retainer, we will consult an expert chemist. Friend Adams at the British Museum comes to mind as a good choice."

THE FOLLOWING MORNING, we duly called at the museum, and were led down to the basement sanctum where Dr. Adams himself worked. But we were greeted only by an assistant.

"Are you from Canvey Island?" the young man asked excitedly on seeing us. His face fell comically on being told that we were not.

"We have just received the most remarkable news," he explained. "You have heard of our famous uranium idol?"

The idol stood on a nearby workbench. Evidently our companion was unaware of our involvement in the discovery of its powers.

"This morning, we received news that a crate had been found at Devil's Point on Canvey Island, apparently washed up by the tide. Its markings indicated it originated from Brazil. Within was found an idol which appeared the twin of this one, except that its features are normal rather than concave.

"I am convinced that if we could bring the two together, they would join seamlessly to make a sphere, proving they were made together, or that one was used as a mold for the other. The original idol was of course found in Africa, so this could be one of the most extraordinary archaeological discoveries of our time."

Holmes nodded thoughtfully. The young man went on, "Unfortunately, Doctor Adams himself is away this morning, but I am trying to have the second head brought here, on my own initiative."

"And where is Doctor Adams?"

"Well, that itself is a rather strange tale. You know that we chemists here have all been baffled by the apparently limitless energy which the idol emits slowly but steadily? This morning there came an invitation from one Professor Challenger of Imperial College: he believed he could explain the matter. Doctor Adams went to see him at once. Can I take a message for his return?"

"No, thank you: we know the Professor, and will call upon them both at Imperial," said Holmes, and we left the young man to his excitement.

As we approached Professor Challenger's study, we could hear the noise of argument within, Summerlee's reedy high voice against Challenger's booming baritone. As we reached the door, Summerlee's voice came clearly: "The trouble is, Challenger, that you cannot tell the difference between a *fictitious* quantity, invented to balance an equation, and physical actuality. Even as some of your *fictitious* exploits rival the tales of the infamous Baron Munchausen."

There came a roar of rage from Challenger, and Sherlock Holmes hastily flung open the door. Challenger, Summerlee, and Adams were seated within. Challenger waved us forward.

"Welcome, Mr. Holmes, Doctor Watson. You are not the students I was expecting, but you have nevertheless come just in time to receive some free education.

"You recall, no doubt, that last time we met I described how we can deduce from relativity that the energy of a moving object rises exponentially as the speed of light is approached. In fact, it departs from the simple Newtonian formula of being proportional to the square of the speed even at low velocity, but the difference becomes significant only at extreme speeds. Similarly, the momentum becomes ever greater as the speed of light is approached. Summerlee and I have been discussing the implications of these facts."

I had no desire to witness the argument restart.

"Surely the dispute could be resolved by simple measurement," I said.

Challenger shook his head. "It is not the quantitative calculations on which we differ. It is the more subtle qualitative point, of how to interpret the results."

I was surprised. "I remember from my schooling that gaining a qualitative understanding of science was quite easy. It was calculating the quantities that made life difficult." I shuddered at the memory of numerical computations and algebra.

Challenger threw out his chest. "Ah, but there comes a third and higher level of scientific understanding, at which the qualitative interpretation is both difficult and vitally important," he said. "Normally, one must first master the mathematics, to be sure. But I feel I can explain the matter so that, exceptionally, you will be permitted to reach the third stage of true enlightenment without the drudgery. I will explain so clearly that the point will be obvious to a dolt."

He glared at Summerlee, but pressed on before that worthy had time to reply.

"What is the physical interpretation of the fact that it becomes harder and harder to accelerate an object, as its velocity approaches lightspeed?

"One approach, which Summerlee endorses, is to assume the mass itself somehow increases: that we must add a 'fictitious mass,' which is a function of the apparent energy of the object—in other words, of the speed of the observer.

"But I am trying to show him," he banged the desk for emphasis, "that it is not a question of fictions. The energy of motion manifests itself as mass, for *mass and energy are in truth one and the same.*"

"Balderdash!" cried Summerlee.

Challenger was unabashed. He picked up a small, ordinary-looking wooden box which had been sitting unnoticed at one side of his desk.

"I have here a subtle device which will prove the point," he said. "Would you care to examine it, Professor?" With great solemnity, he passed the box to Summerlee.

The latter inspected it closely, then pressed the catch to open it. The lid flew up, and a clownish figure on the end of a spring leapt out and struck Summerlee upon the nose, giving him a considerable start. Challenger leaned back in his chair and roared with laughter like a schoolboy.

"Really, Challenger, you must explain this behavior!" Summerlee exclaimed furiously.

"Ah, but I shall. You, Summerlee, perceive an ordinary jack-in-the-box, but the device tells us something quite profound."

He picked up the object. "You see that I can place the jack inside in two ways: either loosely with the spring uncoiled, or with it squashed up under the lid, ready to spring forth. The latter is, of course, the normal operating position."

He closed the box. "Now, Summerlee, suppose the box remains closed. You can of course measure its mass, and tell me, precisely how much energy it will take to accelerate it to a given speed?"

"I dare say."

"Now, does it make any difference whether the spring inside is coiled tight or loose?"

"Of course not: the number of atoms making up the box is unchanged, and so of course is its mass."

"Then, my dear Summerlee, it is that most wonderful of devices: a perpetual motion machine! For we can coil the spring, and accelerate the box very near the speed of light, and open the box aimed so that the jack springs out in the forward direction.

"Now the device will work just as before, for all frames of motion are equivalent. But because objects near the speed of light gain so much energy, the spring will do more work than before to accelerate the jack. We will get more work out of the spring than we put into it, and extra momentum as well!"

Summerlee was obviously flabbergasted. His mouth worked, but he said nothing.

"Do you believe in perpetual motion, Summerlee?" Challenger

pointed to a device on a bookshelf at the side of the room, pictured below.

"I consider it my duty to evaluate inventions brought to me. The occasional wondrous device turns up, but most are flawed. The idea of this gadget is that because the balls on the right are farther from the axle than those on the left, they have greater leverage, like a child cheating on a seesaw. As the wheel turns, the balls run down so that the rightward ones are always farther out. But the device does not, of course, work: a proper analysis shows all forces are in balance.

"And here is a device to break the law of conservation of momentum, by producing a force with no equal and opposite reaction."

He pointed to a device which appeared to comprise several flywheels on a complex rotating mount.

"This item is supposed to produce a lift force in operation. You

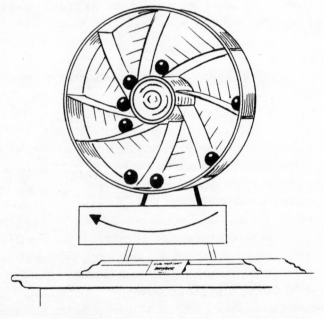

A Perpetual Motion Machine

can place it upon a pair of scales, and as the scales bounce up and down with the vibration, you can imagine the average weight is a little less than before.

"Near us, on Hyde Park corner, men are regularly found proclaiming that such devices work. Would you care to take a soapbox and join them, Summerlee?"

He paused; Summerlee sat silent and immobile.

"But I have an alternative for you. It is that when the spring is coiled, *the total mass of that jack-in-the-box is increased as a result.* Even though not one atom is added, the energy stored in the spring is manifested as an extra mass. So it is a little harder to accelerate the box than before. More force must be exerted, more momentum added: hence more work is done, and there is no perpetual motion device which produces more energy than was put into it. Energy has all the characteristics of inert mass, and I suggest by Ockham's razor that the implication is that the two are one and the same thing in different guises."

Summerlee visibly struggled with himself. "Perhaps you are right, Challenger," he said at last in a hoarse voice. "But really, you should produce additional evidence for such a remarkable proposition."

Challenger beamed. "No sooner said than done, my dear Summerlee!"

He indicated a gadget upon the window ledge. It seemed to be another fallacious device: a little metal paddle wheel, mounted within a glass bulb very like an electric light bulb, so that no current of air or water could possibly reach it.

"It is well known from Maxwell's laws that electromagnetic radiation exerts a force or pressure," said Challenger, "and indeed it is demonstrable by experiment. At the moment, equal sunlight falls on each side of the wheel, and it does not turn. But if I take a condensing glass, so—"

He held up a large magnifying glass close to the window, and adjusted the position so that it focused sunlight on one side of the paddle wheel. The wheel began to turn, slowly at first, then with increasing speed.

The Sunlight Wheel

"Why, that is wonderful," I cried. "A ship becalmed in the Doldrums, near the equator, need only carry auxiliary sails made of reflecting material. Tilt them at the right angle to the Sun, and she will move forward, defying the perversity of the winds!"

"I am afraid not," said Challenger, smiling at my enthusiasm. "The force is in reality quite tiny, by ordinary standards. It is discernible with this device because the bulb contains a vacuum, so there is no air to retard the motion of the wheel."

He went over to a blackboard opposite the window and drew the diagram reproduced on the right.

"This represents a sealed metal tube, which we will suppose is motionless in space, far from Earth, where no forces of any kind—no wind, no gravity—act upon it. It will remain motionless, will it not?"

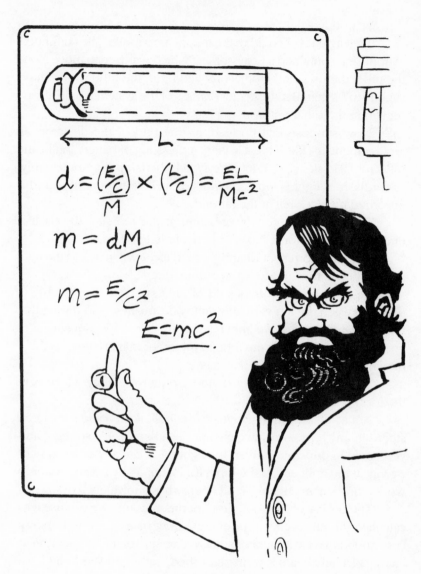

The Remarkable Formula

We nodded.

"Well actually, I could make it appear to move," he continued. "Suppose it contained a heavy object which was free to move along the tube; then when the object moved right, the tube would move slightly to the left. But the center of mass of the system would always remain in the same position, in reality.

"Now suppose that, instead, at one end of the device is an arrangement just like a bicycle lamp, consisting of battery, bulb, and reflector." He sketched it in. "We switch the lamp on momentarily, and a flash of light is emitted, which travels down the tube and is absorbed by black felt at the far end.

"Because light exerts a pressure, at the moment the flash is emitted, the reflector will be pushed a little leftward."

"Ah, but a corresponding rightward force will result as the light is absorbed at the far end," said Summerlee.

"Yes, but because the speed of light is finite, there will be a delay between the leftward and rightward pushes. And during that time, the tube will have moved slightly left. The movement of energy from one end of the tube to the other mimics the effect of a moving mass. Pure energy has mass!"

"Surely you have demonstrated it only for energy which is in the form of light," I said.

"Not at all. The energy to make the bulb flash could have been stored in any form—reactive chemicals, a flywheel, a spring—and you could in principle convert the light back into any other form of energy at the right end of the tube, for example, by boiling water to work a tiny steam engine. So all energy has equivalent mass."

"The beauty of this thought-experiment is that we can use it to calculate the ratio of mass to energy: how many Joules of energy corresponds to one kilogram of mass. The ratio of the momentum to energy of a light beam is well established, both from Maxwell's laws and from empirical experiment. The momentum is just the energy divided by the speed of light."

He scribbled on the blackboard. "Let us write c for the speed of light. Then if the length of the tube is L, the light will be in flight for a time L divided by c. If the mass of the tube is M, it will move at a speed of the light's momentum divided by the tube's mass. So if the energy of the light pulse is E, it will move at speed E divided by c divided by M.

"It therefore moves a distance d, given by . . ." he wrote the formula upon the board. I felt a dizziness coming on.

"I have a horror of algebra!" I cried.

"Ah, but this is really the most profound and beautiful derivation in the whole of physics—and perhaps the most momentous, as I shall shortly indicate. I am sure you will forgive me just this once, Doctor. Now suppose we wanted to move the tube that same distance d by transferring a real mass m from one end to the other. The mass m would be M multiplied by d divided by L. So we can deduce that the mass is equal to the energy divided by the square of the speed of light."

"But the speed of light is enormous," I said. "So the mass associated with a given amount of energy will be quite tiny. This is surely just another of those theological points of no practical significance!"

"You could not be more wrong, Doctor. Let us transfer the c-squared factor to the other side and look at things the other way around."

He wrote the final line upon the blackboard. "The energy equivalent of a given mass is given by multiplying the mass by the square of the speed of light. E equals m times c squared. According to this formula, there is more energy in one kilogram of water—some ten to the seventeen Joules—than in all the coal mines of Wales."

Summerlee snorted. "And you were accusing me of being a perpetual motion merchant. Really, energy from water!" he scoffed. "Even if you are correct, Challenger, no one now upon the Earth has any conceivable idea of how to unravel matter to extract the bound-up energy of which, if I understand you correctly, you are implying it consists."

"Not so, Summerlee. I believe that one of us present here has, albeit unknowingly, been the first man to witness just such an unraveling." Challenger indicated Dr. Adams, who had been struggling to contain himself for some minutes, and now broke in:

"You are correct, Professor Challenger, I am sure of it. I never guessed the solution to my little dilemma would be so profound."

He turned to the rest of us. "You know I have long been troubled by two different features of that mysterious idol. The first was that it seemed to be producing energy from some almost limitless source, albeit slowly. The second was that its atoms of uranium were transmuting themselves into atoms of a different element whose atomic mass was lower. Mass appeared to be disappearing, even taking into account the mass of the strange sub-atomic particles fired out in the process.

"Now both mysteries are explained as one. Evidently," he spoke carefully, "a significant part of what we have always thought of as the mass of an atomic nucleus is in fact a *binding energy*— energy stored up like the coiled spring of that jack-in-the-box in the bonds which hold the electrically charged parts together.

"When the nucleus breaks apart, a positively charged Alpha particle rushes away from the positively charged nucleus, leaving a lighter nucleus behind. The binding energy has turned into kinetic energy of the Alpha particle or, in vulgar parlance, mass has been turned into energy. Sir, you are a genius."

Challenger beamed: modesty was hardly his strong point. "Not merely a genius, sir, but the revealer of forces which will transform the human lot beyond recognition. I am the discoverer of a limitless source of energy."

Adams hesitated. "I fear, sir, that the energy available in principle will still be hard to tap in practice."

We looked at him. "You see, Professor, the actual rate of decay of the uranium in the idol is very small: it would take many thousands of years to halve its mass. And that is despite the fact that the idol appears to be made of some special sort of uranium which decays faster than the substance normally mined.

"We have tried taking small samples from the idol. As soon as the sample is removed from the vicinity of the main mass, its decay rate drops. And any procedure we subsequently subject the sample to—extreme heat, strong acids, electricity—has no measurable effect on that decay rate.

"I fear there seems absolutely no way to increase the rate to the point where useful quantities of energy would be produced."

"That is not surprising," said Summerlee thoughtfully. "After all, the forces which hold the nucleus together must be prodigiously strong, compared with ordinary chemical bonds. Trying to break them with heat is like trying to breach a battleship with a pea-shooter."

Challenger did not appear to be discomfited.

"Ah, but I have a gun of just that potency required," he said.

"Nonsense!" retorted Summerlee. "I would need a little time to calculate the exact energies required, but I am sure—"

"That they are so high, they can be provided only by a similar exploding nucleus," said Challenger simply. "Just as a bicyclist who has freewheeled down from the top of a hill has acquired enough energy to carry him up one of similar height—ignoring friction and air resistance, of course—so the particle spat out by one nucleus has just the energy required to enter another.

"In other words, Summerlee, there is a sort of contagion effect. Given a sufficiently large lump of uranium, so that most particles emitted strike other uranium nuclei before they escape from the mass, the rate of heat emission will rise. The heat can boil water to drive a turbine. The human race will forever be indebted to me, gentlemen, for . . ."

He was beaming at us all from a pinnacle of hubris. Suddenly, Holmes rose from his chair and emitted a ghastly cry. Then he rushed from the room like a maniac.

I caught up with him on Exhibition Road, where he had run into the street and flagged down a hansom. He made no attempt to wait for me, but I managed to scramble aboard.

"A guinea—no, five guineas, if you get us to the British Museum inside ten minutes. Lives depend on it!" he shouted at the astonished cabbie.

Fortunately, the man was quick-witted. He whipped up his horse, and we were soon crossing the Serpentine at a fair gallop, to the amusement of leisurely day-trippers.

I looked at Holmes, who was staring ahead rigid. Fortunately, I have some little experience of dealing with mental seizures. "Come, Holmes, tell me what is the matter," I said in a soothing voice. He looked at me with dilated pupils.

"Do you not see, Watson? The infection factor! The half-globe of the idol is big enough to give a noticeable cross-infection: a fair fraction of the particles the decaying atoms emit trigger more decays. Else the material in the idol would not show more activity than that from small samples taken away from it, as we have just heard Doctor Adams tell us is the case.

"When the two idols are put together to make a globe, the infection factor will rise beyond unity. And then—"

"Pandemic!" I exclaimed.

"Yes, Watson. The plague spreads: more atoms split, and then yet more. But this plague does not have an incubation time of days: it will spread in instants too small to measure. In a fraction of a second, all the atoms will have decayed. A significant part of the original globe's mass will appear as energy. With the force of—"

"A hundred thousand tons of gun cotton!" I said.

Holmes turned to urge the cabbie on, as the horror dawned on me. I looked about with newly opened eyes. In a moment, the great hotels at the far side of the Park might be converted to flying shards of brick, a colossal crater of fire and destruction spreading from the Museum's site at Holborn. The careless revelers who filled the gardens would all be dead or dying.

I could scarcely imagine such a titanic event. The casualties of the greatest battle in history would be as nothing, before the carnage that might start at any second. I am not, I think, a particularly imaginative or nervous man, but I was trembling almost uncontrollably as the hansom pulled up outside the Museum. Holmes did not wait to pay the cabbie, but ran inside. I followed, the man's curses ignored behind me.

In the basement there was no sign of either the idol or the assistant we had seen earlier. Another white-coated man approached us.

"Can I assist you?"

"Yes. Where is the idol, and the young man we spoke to this

morning. It is of vital importance, man!" said Holmes in an authoritative tone.

"They are both gone to Devil's Point, on Canvey Island. My colleague was desperate to test his theory of the idols matching, and there was some delay in dispatching the one washed up to us here. So he took our idol, without the Director's permission I might add, and left a short while ago to catch the train from Fenchurch Street."

Holmes said not a word more, but ran out to the road again. Our cabbie was engaged in vigorous conversation with a Museum guard. Holmes pressed a bunch of sovereigns into his hand.

"As much again, if you can get us to Fenchurch Street for the twelve-ten!"

The man eyed the money, and leapt back on his cab. We rattled along the Euston Road, and ran into Fenchurch Street just as the whistle was sounding for the train's departure. Ignoring the guard's shouts, we scrambled aboard as it moved off.

We were in one of those compartments with no connecting corridor, so there was no chance of searching the other carriages. Holmes looked at his watch and gnawed his lip. "If our quarry is somewhere aboard this train, we are safe, Watson, we will easily intercept him. But what if he caught the previous one?"

I had no answer for him. As the train rattled on, however, doubts began to assail me. The evidence for this relativity nonsense was after all very indirect. Might my companions not be suffering from some more contagious form of *folie à deux*, that madness which afflicts two people together, binding them into a common delusion? Come to that, might Professor Challenger, hardly a responsible man, even have been pulling our legs? My sense of urgency began to diminish.

We were almost the only ones to disembark at the small halt at Canvey Island. There was no sign of the young man we had met that morning. Holmes ran up to the stationmaster.

"Can you tell me, where is Devil's Point?"

The man did not reply, but simply stretched out his arm. From the railway platform, we had a clear view seaward. A long spit of sand ran curving out from the main shoreline for a distance of several kilometers. Some activity was visible at its end. There seemed no way to signal a warning over such a distance.

"Come, Watson, we may yet be in time!"

At that moment, there came a terrible white flash, brighter than the Sun, which temporarily blinded me. We groped about and found each other's arm for support, else we should both have fallen in the massive concussion which washed over us a few seconds later.

My eyesight gradually returned. Through red after-images, I could see a cloud of boiling smoke out at sea. It slowly rose high in the afternoon sky, buoyed up by a narrow whirling column feeding it from beneath. The whole effect grotesquely resembled a giant mushroom or toadstool. As the sea surface became visible, our incredulous eyes beheld only seething water where Devil's Point had lain a minute before. Giant breakers beat furiously on the beach before us. For a moment I thought of tidal waves, but the breakers stopped at the high-water mark. Slowly, the violence diminished.

"WELL, AT LEAST we have found your elephant's foot, Watson!"

I looked at Holmes in bafflement. We were in our rooms at Baker Street the following morning, reading the papers' accounts of the previous day's events. The authorities had moved swiftly to avoid panic. The story concocted, that a munitions boat had gone aground and exploded off Devil's Point, appeared to have been swallowed by Fleet Street—or at least if some editors had doubts, they were wisely keeping them to themselves. There was no mention of anarchists.

"I remarked some time ago, Doctor, that you have an admirable skepticism when it comes to accepting the novel and unlikely-sounding. If there was an elephant in your living room, for example, you might concede its presence only when it trod upon your foot.

"You have been doubtful about all of this new physics, Watson. But now I think even you must concede that we have evidence that not only confirms the reasoning involved but shows that the new discoveries will have significant implications."

"I am still struggling to come to terms with it all, Holmes," I sighed. "It seems such extraordinary consequences flow, from such a small anomaly in the old physics: that the speed of light is constant in all frames of reference."

Holmes smiled. "I have often told you, Watson, that the one singular feature of a case—some small problem which the likes of Lestrade might be tempted to overlook—can be the most important point about it, turning an apparently straightforward story upside-down. Now in the greater world, an apparently simple picture has been similarly overturned: just one apparently inexplicable fact comes to light and behold, all our ideas must change."

"I suppose there may after all be a positive side," I said. "Do you think Professor Challenger's dream of limitless energy for all will now come about?"

"Not quickly, Watson. For one thing the statue is destroyed, and it may prove no simple task to find or make more of that special kind of uranium of which it was constituted.

"There is also a more fundamental problem. You remarked in the context of epidemics, how a tiny change in infectivity can tip a disease from being rare into becoming a pandemic. To release energy controllably from uranium, you would have to arrange it so the infectivity—the probability of one atom's fission causing another's—was very close to unity. Something like point nine nine nine. How close you would be to the brink of an explosion like yesterday's! I would not care to live in the vicinity of such a device. Not until men have learned engineering techniques much more reliable than today's could such a project be sanely contemplated."

I eyed the comfortably familiar surroundings of our lodgings. "Well, talking of sanity, Holmes, my mind has been taken to the very brink, striving to understand what has already occurred. I sincerely hope that whatever else may come our way, this strange new physics has now exhausted its surprises, and I will not be forced to struggle with any further bizarrities."

9

The Case of the Disloyal Servant

"Mysterious vanishing at sea. Stranger than the *Mary Celeste*. Read a-a-ll about it!"

The newspaper seller who barks his wares at the corner of Baker Street is a talented example of his profession. Even if the headlines in the latest edition are dull, he always finds some teasing phrase to shout which makes it difficult to pass him by. Over time, I have developed a certain immunity to his methods. But ever since I read about the strange case of the *Mary Celeste* as a child, maritime mysteries have had a certain hold over me, and today I found myself parted from my coin and eagerly scanning the front page as I ascended the steps of our lodgings.

The story concerned the cutter *Alicia*. The facts were bafflingly simple. One fine morning a week ago, the *Alicia* had sailed into a small patch of sea mist, observed from a distance of some four leagues by her fellow vessel *Sea Eagle*. She never emerged. The *Eagle* had cruised to and fro about the area as the mist evaporated in the noonday sun: there was no sign of the *Alicia*, although a small amount of flotsam was recovered which indisputably came from her decks, even including a labeled life preserver. The *Eagle* had just made port with her strange story.

"I have a curious item here for you, Holmes," I greeted him as I entered our rooms. He glanced up, but did not take the paper I held out to him.

"Ah, you are referring to the mystery of the *Sea Eagle*," he said. "It was already described in this morning's *Times*."

"No, Holmes, the *Times* has its facts wrong. They must have rushed to get an early version of the story in as they went to press. The evening paper here has a more accurate account. You will see the *Alicia* is the ship which has vanished: the *Eagle* was merely an observer."

Holmes nodded. "That is also what the *Times* reports," he said. "But which ship suffered the mystery: the observed, or the observer? After all, a ship lost at sea is a tragedy, but not so unusual. It is the contrasting experience of the *Eagle* which is baffling."

"But this happened suddenly, in calm weather, in clear deep water," I protested. Then the impact of his words struck me. "You are not implying that the captain and entire crew of the *Eagle* are— why, good heavens, Holmes!" I cried.

Holmes smiled. "Fear not, Watson: I am not suggesting that a respectable English merchantman would turn pirate, or her crew take it into their heads to scuttle a fellow vessel. I am merely pointing out that it is the difference between the experiences of the two ships, rather than the fate of either one considered on its own, which is anomalous. Our friend Professor Challenger dropped by while you were out this morning, and gave a plausible explanation.

"He pointed out that waves large enough to swamp and drag down a vessel the size of the *Alicia* have been reliably reported in a variety of circumstances. For example, tsunamis, which are caused by underwater earthquakes, or waves caused by the interaction of tidal flows with channels or ridges on the seabed, or waves which are mere statistical flukes."

"But the *Eagle* experienced no such wave!"

"Challenger's point is that waves are rather unlike material objects in the ways in which they can add together. One apple plus one apple always yields two apples. But consider two waves sharing the same patch of sea. They may appear as two separate waves. But they can also overlap at a given moment, crossing one another's path, so as to appear as a single large wave. And there is a more subtle possibility. Suppose that the peak of one wave coincides with the trough of a similar one. The net effect is that the water level is normal: the waves cancel.

"Challenger claims that two or more waves could pass by or through one another in such a way that the ocean's surface is at one point almost unaffected, while at another point the peaks or troughs add to produce a disturbance quite capable of swamping a ship."

I snorted. "It sounds contrived to me, Holmes. But I can see our news vendor has had me over, as usual: the case is hardly as inexplicable as that of the *Mary Celeste*. Now there was a mystery! But no doubt you deduced the true story of that vessel long ago."

Sherlock Holmes shook his head. "I can think of far too many explanations for that so-called mystery. A ship with a small crew, the captain's family and a few others, is found adrift minus her lifeboat and navigational instruments, but with all provisions and cargo intact. There are at least seven plausible explanations.

"When I heard the story as a boy, it did much to interest me in practical detection. The mystery was difficult in the abstract, but had I been able to examine the vessel, I pondered on the wealth of clues it must have been able to provide. Nowadays I am accustomed to making deductions from a single piece of clothing or personal item associated with a crime. How much more information is to be found on a sailing vessel! The disposition of each rope, each utensil, every piece of equipment present, would have told its own story. I could have read its history like a book, even in those days, given the opportunity."

"You should publish your explanations anyway, Holmes. I have never read even one plausible hypothesis."

Holmes started to fill his pipe. "One explanation that fits the circumstances perfectly involves the fact that her official cargo was listed as industrial alcohol. Now given that—"

At that moment he was interrupted by a thunderous knocking on our door. Holmes rose and walked swiftly to the window.

"A prime clue to the status of an unexpected visitor, Watson, is the vehicle that brings him. A four-horse carriage with the royal insignia hastily concealed by tarpaulins, for example—what does that tell you about our caller's rank and urgency as compared with, say, a hansom cab?" he said calmly.

I advanced and opened the door. A tall man of military bearing, who I fancied looked a trifle uncomfortable in civilian clothes, stood on the threshold. I ushered him in.

He greeted us with a stiff bow. "I am Captain James Falkirk of the Household Guards."

Holmes raised his eyebrows. "Are you bringing us a royal summons?"

The captain eyed him intently. "Not as such: I am here on my own initiative. Sirs, will you each state upon your honor that you are loyal citizens of the Crown, and will be bound by absolute discretion as regards what I have to say?"

I was outraged that he should think such a request necessary, but Sherlock Holmes assented immediately, and I followed suit.

"There has been a tragic occurrence at the Palace. A violent death has occurred," said the Captain.

I could not contain a gasp. "You do not mean—"

He shook his head. "Neither one of the family, nor a guest of status. The dead man was a mere stableman, and one whose recent actions were about to bring disgrace upon even his low station. Frankly, he will be little missed.

"The man—Jenkins by name—had been in our employ for some ten years. Last year a scullery-maid was dismissed, having got herself into trouble of a type which you, being men of the world, will understand. She refused to identify the man involved. It has now emerged that Jenkins was the father.

"It appears that he had some rudimentary feelings of responsibility toward the girl, and attempted to pay for her support. Being of very modest means, but having some contacts upon the turf, he attempted to increase his wages by gambling. The attempt turned sour, and soon he was at his wits' end.

"It was then that a plan came to his low mind so wicked that you will think it almost inconceivable. He went in turn to three of the tawdriest rags in Fleet Street, offering to sell stories of life at the Palace to their editors. He even hinted at knowledge of some minor scandals.

"This morning, however, his conscience caught up with him, and he decided he could not carry through with his scheme. He came to me in tears and confessed all. He begged my forgiveness. This of course I withheld, but I agreed that his mistress and her child would be taken care of, in view of his repentance."

"A pity that such a compassionate solution had not occurred to you before," said Holmes dryly. "A little enlightened self-interest earlier in the day might have averted problems all around."

"I could hardly have condoned their conduct in such a way. However I now pledged what the man requested, although I felt the situation was close to one of blackmail. He himself would of course be dismissed, and I made it clear he would never be given a reference, nor find any respectable employ, once having worked out his notice. This afternoon he went into a room adjoining the stables and shot himself through the head."

"A safe ending, from your point of view," said Holmes coldly. "And will you now keep your pledge as regards the girl?"

"Sir, I am a man of my word! She will be transferred to a distant royal household—Balmoral comes to mind—with a story that she is a widow, which will now be close to the truth. But the matter has not in fact ended safely, for there is a small puzzle about the death, which, if misconstrued, could yet cause a scandal the like of which a Royal household has never seen."

Holmes leaned forward with interest. "Pray explain the problem."

"The man was inconsiderate even in the manner of his death!" said Falkirk indignantly. "First, he used a rare and unique make of gun, a German device from the Prince's private collection, given him by his Prussian cousin. Second, he committed the deed mere feet away from the verandah where the Queen and her co-hosts had assembled to greet guests to this afternoon's Royal garden party."

"His action can hardly have remained a secret, then," I said.

"It did, for the gun he took was an almost silent air rifle. No guest appears to have heard the report, and even of those on the verandah only the Queen and her guest the King of Molstein claim to have heard the shot."

"Claim to have heard?" Holmes said sharply.

"That is the strange thing: two other guests on the verandah, who were equally close to the shot, heard nothing. You see the implication. The stableman's story, if he had really sold it, could have caused the family great embarrassment. It might conceivably be alleged that the man did not commit suicide at all but was murdered, the Queen and her royal guest claiming to have heard the shot to establish a false time of death. Thus enabling an alibi for the murderer to be established.

"It would be manifest nonsense, of course," he added emphatically. "The idea that a royal retainer would commit murder, and the Sovereign then protect him from justice, is inconceivable!"

I could not help thinking that Thomas à Becket would have been surprised to hear it, although that episode was admittedly some centuries ago.

"But the mere shadow of suspicion could do inconceivable harm to the Family," Falkirk continued. "A discreet investigation, to establish the facts beyond all doubt before any public statement must be given, would be invaluable."

"I quite understand, and we shall come at once," said Holmes.

The captain sprang to his feet, relief on his face. "Excellent! The carriage awaits us downstairs."

Holmes shook his head. "Hardly compatible with discretion, that I should ride to the Palace in so thinly disguised a vehicle. No, you return in it, and we will proceed by more humble means to meet you shortly."

The captain nodded, then hesitated on the threshold. "If the carriage is so conspicuous, my journey here will already have been noted."

"Have no fear. That might well denote a private problem of your own, not involving the Palace at all. If I am asked, I shall merely say that you have got a scullery-maid into trouble, and are seeking my advice to hush the matter up." Holmes smiled at our visitor's horrified face.

A HALF HOUR later, we were admitted to the Palace by a discreet side door. We were led through servants' corridors for some distance, then into a long room whose only light was from two narrow slit windows at one end. At the opposite end, huddled along the wall, lay the body of the unfortunate stableman, still clutching a rifle of curious design. The floor was covered by a thick red pile carpet, the walls by tapestries: a man whose life had been led in the Palace's humblest servants' quarters had entered one of the royal residence rooms to end it.

Holmes knelt by the body. He pulled the rifle gently from the dead man's grasp. "I recognize the work of the blind German gunsmith Von Herder. A very discreet weapon. I shall demonstrate."

He checked that the breech was empty of bullets, then pumped the rifle up and pulled the trigger. A sound like a brief, deep blast from a motor horn was the only response. Holmes cocked his head.

"An almost pure note of about one hundred and sixty-five cycles per second. The sound is almost like an organ pipe, the device is so beautifully made. Certainly less conspicuous than an ordinary silenced gunshot, yet quite audible through these windows if they were open." He walked over to examine the windows, which were equipped with heavy burglar-proof shutters. Both were opened wide. "Were they each like this at the time of the shot?"

"It seems probable," Falkirk answered. "The body was discovered by a maid who had entered to dust, and may possibly have adjusted one or both windows before seeing the corpse. She is still quite hysterical, and it is difficult to get any coherent information from her."

"Very good. Let us now examine the terrace beyond," said Holmes. Falkirk led us through an adjacent room and out onto a raised terrace, balustraded to make a wide balcony, which overlooked a beautifully manicured lawn studded with flowerbeds.

"At the start of a Royal garden party, the hosts initially come out upon this verandah to greet the guests," explained Falkirk. "Today there were four official hosts: in order of rank the Queen, her senior guest the King of Molstein, the Archbishop of York, and Sir Oswald Launton."

Holmes smiled. "A Knight, a Bishop, a King and a Queen. Placed in jeopardy by a man whose status made him a mere Pawn. I have always had a fondness for chess puzzles. Can you show me just where each was standing, when the shot was heard?"

The captain looked shocked at this levity but complied. He indicated four positions on the flagstones, which I reproduce in the whimsical but spatially accurate sketch overleaf. "All four stood just behind the balustrade. The Queen occupied the central position, with Sir Oswald to her right and the Bishop to her left. The King stood to the left of the Bishop. If both windows were indeed open, all four should have heard the shot clearly, one would think."

"Of course, one window may have been shut," said Holmes thoughtfully. "If the right was closed, for example, the Knight would have been farthest from the sound; if the left, the King might not have heard it. Yet in actuality, both the Queen and the King heard the sound clearly, but the Knight and the Bishop heard nothing. It is certainly a curious circumstance."

Suddenly a light came into his eyes. He produced from somewhere on his person a tailor's measure, and proceeded to ascertain the separation of the slit windows, which appeared to be about three meters, and the distance from the wall to the balustrade, which was four meters.

"He looks like a suspicious buyer inspecting a house," commented Falkirk quietly. "I am afraid he will find that this property is not for sale."

At that moment, Holmes straightened up with a cry of triumph. "The note sounded was one hundred and sixty-five cycles per second," he said confidently. "You are aware that sound is a wave, Watson, and that the waves travel at some three hundred and thirty meters per second. So the wavelength of the note would have been?"

"Two meters," I said briskly. There are times when I feel that Holmes underestimates me.

"Now, Watson, earlier today we were discussing how water waves can cancel one another's effect, if the peak of one coincides with the trough of another. Sound waves are of a slightly different

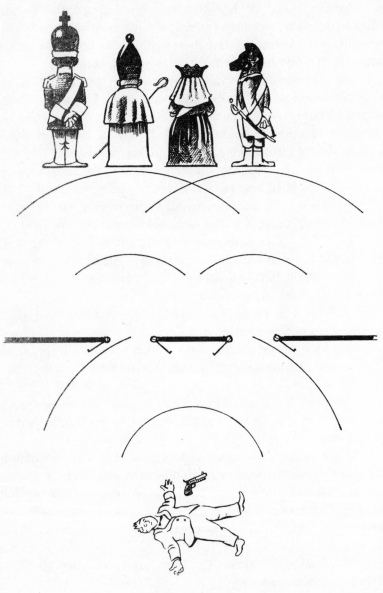

The Royal Party

nature to water waves: they are alternating zones of compression and rarefaction of the air. In this case, the gap between successive fronts of maximum pressure—or equally, successive fronts of minimum pressure—was two meters. Does that suggest anything to you?"

"Well, I suppose two sound waves might possibly cancel, if the compression zone of one coincided with the rarefaction zone of another," I said. "But how you would determine when and where this might occur is well beyond me."

Holmes smiled. "It is simplicity itself, my dear Doctor," he said. "The Queen was standing at a point equidistant between the two windows: wave peaks emerging from both at the same time would thus reach her simultaneously, and the effect would reinforce rather than cancel.

"The Knight, on the other hand, was four meters before the right window, two wavelengths, but five from the left, a messy two-and-a-half wavelengths. Peaks from the right window would have coincided with troughs from the left, and vice versa, canceling the sound exactly. Behold, the Knight hears nothing!"

Our host held an expression of dawning relief.

"The Bishop is four meters before the left window, five from the right, and hears nothing also, for just the same reason as the Knight. The King's position is more interesting. He is some four meters from the left window and six from the right. As a result, consecutive peaks of pressure reach him together. Their effect is to reinforce one another, and he also hears the sound clearly.

"You can rest easy, Sir Oswald,"—our guide and I started together—"yes, of course I saw through your alias. There is no need to doubt your Sovereign's word, or to consider whether you must jeopardize your honor by lying to protect her. You and the Bishop heard nothing, the Queen and the King heard the shot clearly, and there is no contradiction involved."

"YOU ARE TO be congratulated, Holmes, on solving a puzzle well outside your usual field," I remarked as we strolled back northward along the pleasant boulevard in Hyde Park opposite the Park Lane hotels.

Holmes shook his head. "Not to as great an extent as you think, Watson. When Challenger called by this morning, he described a scientific experiment to me of a very similar sort. In fact, he believes this phenomenon of a wave interfering with itself to be a most important one, which will give him the upper hand in his long quarrel with Professor Summerlee as to whether light is truly a wave. His purpose in calling was to invite me to a public debate with Summerlee, which starts at five o'clock today. Evidently he wishes persons whose intellect he respects to be there, to witness his anticipated triumph."

He glanced at his pocket watch. "The debate must be starting now. I had not really intended to go. But I suppose we owe him his due for helping us with our case, however coincidentally. It is at the Royal Society's lecture theater in Burlington Place, close by here. Shall we make a small detour?"

We entered the auditorium to the sound of Summerlee's high, reedy voice. Although we tried to take our seats inconspicuously, the hall was of that type where the seating slopes steeply upward from the front, and we were in clear view of the podium. Challenger was seated to the right of the lectern, sprawled insolently at ease; his eyes turned up as we sat down and he gave us a heavy-lidded wink.

"So it has been shown that there is no such thing as an ether to transmit electromagnetic waves," Summerlee was saying. I nodded: the speed-of-light experiments I was by now quite an expert on had demonstrated this quite conclusively, I felt.

"But I do not depend on such theoretical arguments to show that my point of view is correct," Summerlee continued. "There are two quite different experiments, each easily performed, which both show clearly that light is made up of discrete particles, or photons as I call them, just as all matter is composed of discrete atoms.

"First, consider what happens when light is shone upon an appropriately prepared metal surface. The absorbed energy causes electrons to be emitted from the surface, and the number of such electrons, and their velocity, can readily be measured.

"What happens if we double the intensity of light falling upon the surface, keeping the color unchanged? By my colleague's wave theory, you might intuitively expect that both the number and velocity of the electrons would rise. In fact, we observe that the velocity of the electrons is perfectly unchanged, but their number doubles, as the number of incident photons has doubled.

"More revealing is what happens when we change not the intensity, but the color, of the incident light. For example, we could double the temperature of the bulb filament, so that the hue changes from dull red to blue-white. Challenger would have you believe that the color change is due to a halving of the wavelength of the light. But I tell you that it is a doubling in the energy of each individual photon—each particle of light—that is emitted. I know I am correct, because if I shine the same total energy—the same wattage—of light as before on the metal, but of the color blue rather than red, the number of electrons emitted halves, but the energy of each doubles. This is easily explained if each single photon knocks out one single electron: there are half as many photons as before, each of double the energy. But it is almost impossible to explain the outcome in terms of waves.

"The second experiment is still more conclusive. It involves reflecting light of a quite different color—invisible light in the form of X ray radiation—off certain crystals. You will be aware that normally, of course, reflected light is of the same color as incident light. But is that always so?"

He produced a small indiarubber ball from his pocket and threw it at the back wall of the lecture theater, catching it as it rebounded. There was a brief handclap from some students toward the rear of the hall. Summerlee frowned at them.

"Supposing that that ball were perfectly elastic. Would it return to my hand at the same speed it was thrown?"

A hand went up, and Summerlee nodded to its owner. A tentative voice spoke: "Slower, sir. The wall is not infinitely massive, nor infinitely stiff, and it will recoil to a small extent, so robbing the ball of some energy."

Summerlee nodded approval. He took a football from beside the lectern and handed it to Challenger, who raised his shaggy eyebrows. "Sir, in your own time, would you be so good as to toss this a few feet vertically into the air."

Summerlee walked to the far side of the platform. Challenger shrugged, but tossed the football as requested. At the apex of its flight, Summerlee hurled the indiarubber ball with great force, striking the football at its exact center. For a moment, we could all see a triumphant schoolboy bowler suddenly transported fifty years in time. The football was noticeably affected by the impact, and fell to one side as the indiarubber ball returned to Summerlee's hand more slowly than it had started. There was a wider burst of applause, in which Challenger ironically joined.

"I have just demonstrated," said Summerlee severely, "what happens when X rays strike a crystal as I described. The X rays return with reduced energy, as if they had struck an object which had recoiled.

"Electromagnetic radiation carries a calculable momentum, but one that is far too small to make the whole crystal, or even the layer of atoms making up one face, recoil significantly. However, if we understand the interaction as individual photons striking individual electrons from the crystal, then the energy loss measured in the X rays is exactly that to be expected as individual electrons are knocked backward. So the radiation is not a wave: it is a hail of photons, each of which strikes a particular electron and rebounds with reduced force as the electron recoils like the football."

There was another wave of applause, this time coming as much from the gray-bearded men in the front row as from the young students making up the remainder of the audience. It seemed to me that Summerlee had won the debate conclusively. But as he sat down, Challenger rose majestically and beamed around quite unperturbed.

"Ladies, gentlemen," he said, "my colleague has given you a most plausible account. He has demonstrated that in certain *interactions* between light and matter, light energy is absorbed or reflected in discrete packets, whose energy varies with the color—or in my terms, the wavelength or frequency—of the light involved. The inference that light must therefore consist of discrete packets is a tempting one, which might even have deceived an intelligent man." He beamed around, as Summerlee sat rigid.

"But it does not follow at all. If I may pick a homely analogy, familiar no doubt to the students in the audience: let us go to a pub, and observe the landlord as he draws beer from the barrel for his customers. We see that it is invariably drawn out in exact multiples of one pint. Do we therefore conclude that the beer consists of solid indivisible lumps of matter of that volume? No, it is a fluid which may in fact be divided in much finer proportions."

"You can't have visited our neighborhood tavern!" called a youthful voice from the audience. There was an ironic cheer, which the Professor ignored.

"We will therefore not bother to refute, but merely ignore, Professor Summerlee's utterances. For I can show you unarguable proof that light is in fact a wave, spread out through a continuous volume of space." He indicated some apparatus in the well of the theater.

I craned my head forward. In front of the platform stood a pair of billiard tables, as shown overleaf. One was normal except for some wooden dividers which had been set upon its surface. The other looked identical, but shimmered oddly: on closer inspection, the surface was covered in a couple of inches of coloured water. The overall effect was distinctly reminiscent of a college bar on the night after Finals had finished. Challenger moved behind the dry table, and picked up an ordinary billiard cue. A large number of balls were gathered at his end of the table.

"We shall suppose that the balls are hit at random. So I shall invite Professor Summerlee to do the honors. Sir, would you please strike these balls toward the far end of the table?" He indicated two holes in the central partition, through which the balls would have to be hit accurately to pass.

Summerlee came forward willingly, and struck balls in rapid succession as Challenger passed them to him. But whether Summerlee's coordination had failed him, or the balls or cue been sabotaged in some way, they seemed to leave in purely random directions. I was reminded of a humorous curse on billiards cheats in a new Gilbert and Sullivan opera I had attended, that they should be made to play:

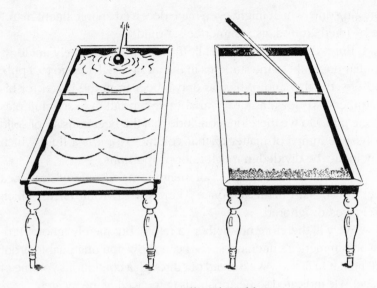

The Two Billiards Tables

On a cloth untrue
With a twisted cue
And elliptical billiard balls

Most balls struck the central partition, and Challenger quickly removed these from the table. Some did pass through the holes, but even these were deflected at random as they ran through sticky puddles of blue and red paint which had been deposited there. The result was that the corridors at the far end of the table filled up randomly, with an approximately equal number of balls ultimately at rest in each.

Challenger stepped forward as Summerlee hit the last ball: "Unknown to himself, Summerlee has been simulating a very old experiment in which light is emitted from a source and allowed to pass through two slits. He has been reproducing it according to his photon theory of light. You will observe that there are now about the same number of balls, or photons, in each end pocket.

"I ask you now to consider what would have happened if I had blocked one gate—for example, that marked with blue paint. The red spotted balls would be present just as they are now; the blue would be absent. Obviously, the total number of balls in each individual pocket would then be less than, or at most equal to, those you see now. *Closing* one of the gates could not possibly result in *more* balls in any pocket, whatever tricks I have played with the design of the balls or the table.

"Now let me show you an alternative model." He moved to the water-covered table and activated a small mechanism at the end, which started to bob up and down emitting a pattern of ripples. The ripples spread and passed through both gates, and on to the table end, causing the water level at the far end to oscillate up and down in a curious pattern. The end was blocked with wax paper, so that water did not actually escape, but the paper became dark to the local high-water mark. At regular intervals, the dark strip reached a height of some two centimeters, but it diminished to vanish altogether at positions halfway between these peaks.

I felt a revelation come to me. "Why, it is just like the layout of the murder at the Palace, Holmes! The bobbing object represents the gun, or source of sound; the central barrier is the wall with its two windows."

My friend nodded. "Indeed, if we were to place chess pieces at the points where the end waves are strongest and weakest, the model would be perfect," he replied quietly.

Challenger had the air of a conjurer about to perform the final stage of a difficult trick. "Now watch very carefully," he said, "as I close one of the gates. Look at a point of the end wall where the water is at present stationary."

As he closed the gate, the pattern on the end wall vanished. Instead, the water level across the whole end now rose and fell as the ripples came through the one gate still open.

"You see that *closing* a gate has *increased* the impact at previously quiescent points," said Challenger, "something which could not possibly happen in the case of the billiard balls."

He walked to the back of the hall, where several large photographic plates stood veiled. "The identical experiment has often been done with light. When light passes through a pair of slits, it produces a similar pattern of cancellations. In technical language, the waves interfere with one another.

"To clinch the matter, with the aid of a very sensitive photographic film devised by my friend Doctor Adams, I repeated this experiment with a source of light so faint that, by Summerlee's reckoning, only a single photon at most was present in the apparatus at any one time. This eliminates the possibility that the photons might somehow jostle or affect one another to produce a wavelike pattern.

"First I performed the experiment with a single slit open." He unveiled the leftmost photographic plate: it was a uniform gray.

"As you see, the light was spread out uniformly over a wide angle. Then I repeated the experiment with both slits open." He unveiled the second plate to reveal a strong pattern of light and dark bands. "A clear interference pattern is produced. In Professor Summerlee's terms, each single photon would somehow have had to be aware of *both* slits, to produce this pattern. For no photons at all have struck certain points on the plate, which are characterized by the fact that their distance from each of the two slits differs respectively by exactly one-half wavelength of the light."

He paused as a new ripple of applause ran round the hall, which he held up his hand to quell. "You see that, whatever its behavior when it interacts with matter, light is in its own true essence a wave. But this experiment was first done at the start of this century, and its implications have as long been obvious to clear-thinking men." He glared at Summerlee. "It is not to go over this old ground that I have agreed to be here today. It is to announce a quite novel finding, which, I venture to say, will shake the world of science to its foundations."

He paused to a breathless hush. It struck me that, were it not for his scientific genius, he could have made a successful career as an actor or stage orator.

"It occurred to me that it would be interesting to perform the two-slit experiment with particles rather than light. It is nowadays possible, with suitable apparatus, to produce a stream of electrons, or atoms, or even molecules, emitted individually at a known rate. Doctor Adams's film is sensitive enough to record the impact of such individual particles, just as if they were Summerlee's 'photons' of light. I must confess that I expected this time to see a result like that with the billiard balls on the table before me. After all, unlike a wave, each electron must pass through one slit or the other only. To avoid the possibility of electrons jostling one another to produce wavelike behavior, the current in the apparatus was kept low, so that only one electron was in flight at any given moment.

"With a single slit open, the electrons scattered randomly, and the plate was uniformly fogged, as anticipated. With two slits open, I confidently expected the same uniform fogging, but of twice the brightness. This is what I found."

He unveiled the third plate. A collective gasp arose from the audience. There was a clear pattern of light and dark stripes, just like that produced by the photons.

"It is well known from many experiments that electrons are tiny, indivisible particles. But this effect shows that each electron must somehow have passed through *both* slits. The only conclusion is that the electron is really a wave. Its presence at a given unique point must be illusory.

"I repeated the experiment with atoms, and then with whole molecules. In every case, I was able to produce the same pattern of wavelike behavior. Ladies and gentlemen, I have now proved that solid matter of all kinds is mere illusion. The Universe is composed entirely of waves. Why, even you, sir"—he turned to his fellow Professor—"are not in truth a solid object, but a mere package of waves, slowly dissolving at the edges with the passage of time, a little local disturbance upon a boundless sea." He turned his back on Summerlee and bowed solemnly to the audience left, right, and center.

The auditorium was silent for a moment, then a thunderous wave of applause broke. However, I noticed that many of the professional scientists present did not join in wholeheartedly, but were frowning thoughtfully as if they struggled to grasp the concept. Nevertheless, no one seemed disposed to argue with the redoubtable Professor, and as people started to rise and stream from the hall, Holmes and I joined the flood.

10

The Case of the Deserted Beach

THE WINDOWPANES RATTLED AS THE first puffs of the first gale of autumn beat against them. I sat alone in our Baker Street living room. Sherlock Holmes had been out for most of the last week on the track of some case of which he had told me little, and I was feeling a touch neglected. The noise of approaching footsteps on the stair was therefore welcome; but it turned out to be only the telegraph boy. The red Express envelope was addressed to Mr. Holmes, but I opened it, as I had his standing permission to do.

The contents were terse: BODY FOUND ON BOURNEMOUTH BEACH IN APPARENTLY IMPOSSIBLE CIRCUMSTANCES STOP EVIDENCE ABOUT TO BE DESTROYED BY TIDE STOP PLEASE COME IMMEDIATELY IF FEASIBLE STOP SIGNED GREGORY.

I hesitated. I had no idea of Holmes's current whereabouts, and delay would obviously hinder any help we could give. Yet I knew that Holmes had high hopes of Inspector Gregory, a provincial detective who was expected to apply to Scotland Yard as soon as his case record had built up to a sufficient level.

In the past, I am bound to admit I have had very indifferent success when I have tried to investigate a case without Holmes's assistance. But this problem obviously involved forensic evidence, which my medical training would assist me to gather, and if I could give my colleague a personal eyewitness account it might provide him with vital clues. I consulted our volume of railway timetables, then scribbled a reply on the form the boy presented to me, asking Gregory to meet the five-fifteen train. Leaving a note for Holmes to follow me if possible, I wrapped up thoroughly against the chill, and embarked for Victoria Station.

I was some thirty minutes early for the train, and I boarded the end compartment of the rearmost carriage, nearest the station entrance, in the hope that Holmes might catch up before we departed. To my relief, just as the guard's whistle sounded its second blast I heard hurrying feet, and the door next to me was flung open. It was not my colleague who scrambled aboard as the train jerked forward, however, but the bulky figure of Professor Challenger. He seemed as surprised to see me as I was him.

"Good afternoon, Doctor. Taking a break from your practice?" he wheezed as he recovered his breath.

"In a sense," I replied guardedly, for Holmes had impressed upon me that police requests to him were always to be treated with the utmost confidentiality. "And yourself?"

"Why, in a sense, Doctor, I am continuing our little adventure in the North Sea." He wagged a finger at me playfully. "I am sure you remember that the solution to an apparent paradox of motion lay in a hidden wave—one which operated out of sight, yet had a very real effect on objects upon the surface. That episode was most instructive. It provided a powerful metaphor; it helped to inspire my discovery that all the contents of our Universe, solid matter as well as light, are in a true sense waves."

I remembered his recent demonstration with the billiards tables. "Yes, your view would seem finally to have prevailed over Summerlee's," I said as the train gathered speed.

Challenger shrugged. "Summerlee did not concede so readily. The evidence for the particle-like nature of atoms is rather convincing—atoms, unlike his alleged photons of light, can very definitely be observed to be in a well-defined place, at a given time.

"With my study of electrons fired through two slits, I was able not only to show that they behaved in a wavelike manner qualitatively but also to deduce the specific wavelength involved. This wavelength turned out not to be constant but to vary with the electrons' speed. The slower the electron, the longer the wavelength. The same held true for other particles. But Summerlee was able, in many different ways, to establish that the size of the electron was quite tiny—so far, immeasurably small, certainly far smaller than the wavelengths I had demonstrated. As we both repeatedly refined our experiments, my wavelengths grew ever longer, his electron-particle ever tinier, in seemingly hopeless contradiction.

"We finally agreed on an experiment we would both regard as definitive. Summerlee had found an electric probe so sensitive it could detect the passage of a single electron. We decided to set up my two-slit experiment with one of Summerlee's probes by each slit. We should unambiguously see whether individual electrons passed through one slit only, or somehow split and passed through both."

He shook his head gravely. "I think it would be fair to say that the result surprised us both equally. We did a dry run with Summerlee's electrodes switched off. The usual banded pattern of interference appeared on the photographic plate. Then we switched the probes on for a second run. As Summerlee claimed, each electron passed unambiguously through one or the other slit, lighting an indicator lamp accordingly. But when we developed the photographic plate, the banded pattern of interference had disappeared!

"It really seemed as though the Universe was in some kind of conspiracy to present two Janus faces of reality to different experimenters. Electrons behave as waves, until you try to detect them as particles—whereupon the wave-like behavior vanishes! This was the dilemma that led to Summerlee's unfortunate breakdown."

I had heard nothing of this, and my face must have betrayed my surprise.

"The poor fellow has declared that he does now believe in waves—but only waves of *probability*." Challenger tapped his massive head significantly. "He claims that the position of an atom, an electron, or whatever, can be described only as a distribution of probabilities, until it is measured. The mere action of measurement then somehow causes the particle to pop into existence in a definite place within the probability field."

I felt Challenger might be giving a rather biased view.

"Well, after all, an electron is a rather abstract entity, at least to us large creatures, is it not?" I said. "I would be quite comfortable to consider the thing a mere confluence of probabilities."

Challenger shook his head vigorously. "The effect can be demonstrated for atoms, for molecules, and in principle for still larger things," he said. "Why, in Summerlee's view you could take a cat, or an elephant, or—why, let us apply poetic justice, and take Summerlee himself."

A schoolboy gleam came into his eye. "You have seen that circus act where a performer is fired out of a cannon—which is really just a sort of catapult, of course, with a little flash-powder to make a more impressive effect—and flung across the tent to a net on the other side?" I nodded. "Well, let us take Professor Summerlee, and place him in such a cannon. We will fire him a great distance, in the general direction of a wall in which two slots are cut. We repeat the experiment over and over. In a percentage of cases, Summerlee will pass through one slot or the other, and come to rest on the far side of the wall."

Challenger raised a finger. "Now we will make a small change. We will perform the experiment as before, but this time in pitch darkness. As his trajectory is no longer observed, wave-like behavior will occur, and Summerlee's landing points will presumably begin to build up an interference pattern. If we place a sand pit for him to land in, like that provided for long-jumpers, and do not rake it between trials, a regular pattern of hollows will be carved out in due course. Of course, Professor Summerlee's theory of probability waves can explain this."

"It sounds a difficult experiment to perform, not least for the problem of obtaining the cooperation of the subject," I said.

"But possible in principle, nevertheless," said Challenger unabashedly. "Now suppose we ask Professor Summerlee, after a particular landing in the sand pit, what he has just experienced. You are a practical man, Doctor. Would you really expect him to say: 'I felt myself diffuse into a ghost-like cloud of probabilities as I left the catapult. Some of me impacted the wall, some of me passed through each of the slits, but it was not until you turned on the light to inspect the sand pit that I felt myself coalesce in this spot just now'?"

I was forced to admit that put that way, it did sound rather absurd.

Challenger snorted. "It is bunkum, Doctor. Sheer hand waving and philosophical sophistry, marking the ruin of a mind which, if never great, at least once had some semblance of competence."

I sought to prick his arrogance somewhat. "No doubt you have a complete explanation for these baffling results, then?" I said dryly.

Challenger beamed at me, my sarcasm evidently unrecognized. "I mentioned that I was, in a sense, about to continue our North Sea adventure. Whereas Summerlee is a theorizer, I am primarily a man of action. To better understand the nature of waves, I have set up a marine Wave Research Station outside Bournemouth, with a small staff. Water waves are of course different in character from their more abstract mathematical cousins, yet there are striking parallels. Some skepticism from Summerlee notwithstanding, the station has already yielded useful insights. I am now about to announce quite a breakthrough, I think."

With that he fell silent. After several unsuccessful attempts to draw him out further, I turned to watching the scenery pass by, and in due course must have dozed off.

I awoke with a start to Challenger's vigorous shaking of my shoulder. For a moment, I thought we had arrived at our destination, for the train was stationary; but in fact we were halted alongside a bleak stretch of seafront. Large breakers thundered in through a line of rocks, running ashore onto a beach of coarse shingle. It was a most uninviting sight, and caused me to wonder why the English rush lemming-like to the seafront on every public holiday. If our island were situated a couple of thousand miles southward, the habit would be more understandable.

"There, Doctor! Already you see how waves upon water give us an insight into the deeper nature of reality." Challenger pointed. "You see those rocks are on average just a couple of meters apart?

Yet there is a wave fully thirty meters wide, and it passes through the line almost undiminished."

"Well, obviously, Professor: the wave squeezes through the gaps, and re-forms on the other side."

"But the feat would be impossible if the wave were a solid object. And this reveals the secret of two phenomena which have been baffling the best scientific minds," Challenger said, leaning forward as the train renewed its motion with a jerk. The Bournemouth station platform slid into sight.

"There was previously great difficulty understanding how metals, or indeed any substance, can conduct electricity readily. If electrons obeyed Newton's laws of motion, bouncing from each other and from atoms like tiny billiard balls, each could move only a short distance before losing its energy to successive collisions, and all materials would be insulators.

"A second problem emerged with the need to construct lightweight vessels to contain the recently discovered gas helium. If this substance is placed in a thin-walled container, even one which is guaranteed absolutely free from holes or imperfections, it leaks out through the apparently solid wall, like a prisoner tunneling to his escape.

"Both phenomena can easily be understood if electrons and atoms are really waves. For waves can freely pass through an array of obstacles, or through one another, with comparatively little effect."

The train halted with a loud hiss of escaping steam as the engine valves were opened. Challenger swung the door to our compartment ajar and prepared to alight.

"But that could very well be explained in terms of solid particles, and Summerlee's probability waves," I said. "If all locations are a mere matter of probabilities, then a particle close to one side of a barrier might just happen to emerge on the other. A prisoner should press himself up against the bars of his cell door, by this prescription, for with a little luck—lo!—he may find himself spontaneously on the far side."

I had spoken in jest, but Challenger gave a contemptuous snort. With a brusque wave of his hand, he hurried off.

The station appeared deserted: the resort's summer season was evidently quite over. But as I neared the exit, Inspector Gregory came into sight. On seeing that I was alone, his face fell to an extent that was almost comical.

"I was unable to locate Holmes immediately," I explained, "but with luck, he may be following on the next train."

Gregory sighed. "That will be too late for the tide," he said. "In fact, we will be just in time as it is. Our local medical examiner is very competent, Doctor, but still, a second opinion does no harm, and no doubt Mr. Holmes is better accustomed to your eyewitness reports than any others. Let us make our way to the beach."

A police carriage took us down to the shore, and a mile or so along the sea-front, to near the end of the promenade. We descended slime-coated steps set in the sea-wall, and walked over shingles to stand by an expanse of wet sand separating us from the advancing breakers. Gregory pointed to a body lying about thirty meters to seaward. It was surrounded by unmarked sand, except for a narrow row of well-trodden footprints connecting it to the spot where we were standing. We made our way gingerly along this route, and I knelt by the body. It was that of a young man, tall but somewhat flabby muscled. The cause of death was obvious: a mass of clotted blood in his hair marked the spot where he had been struck with great force by some sharp-edged object. The body was wearing a bathing costume which was still slightly damp, but there were no signs of water inhalation or other visible damage.

"It looks like a clear case of murder," I said, rather puzzled as to why Gregory should think the matter obscure. "He was hit a vigorous blow with some object having a fairly sharp edge. Neither a club nor a knife: something more like a rowing oar, perhaps, considering our location. Death must have followed very rapidly, probably within seconds, or there would be other indications such as bruising."

Gregory shook his head thoughtfully. "I would agree with you, Doctor, but for one singular circumstance. The body was spotted some four hours ago, by a pair of officers patrolling the promenade. When they approached the corpse, it lay some twenty meters from the receding sea and was surrounded by damp sand *which was completely unmarked.* No one could possibly have been near the spot, or their tracks would be clearly visible."

I pondered a moment. "What if the murderer was careful to tread in his victim's footprints, retreating backward in the same manner?" I suggested.

"There were no tracks at all, Doctor, not even the victim's. For all signs to the contrary, he might have been dropped from the sky! The obvious possibility, that the body was washed ashore as the tide receded, is ruled out both because the waves would have formed a cavity in the sand around its resting place—you see, we are familiar with the appearance of washed-up bodies, human and animal, in these parts—and also as there would then be water in the lungs, as well as dissolving of the blood clot by salt water: there is no trace of either."

"Perhaps the man could have swum ashore after falling from a boat, and then being struck by the keel, or by an oar as I first suggested?" I offered tentatively.

"No, he would have been unconscious in seconds: no boat could have been so close inshore without running aground, in these waves." He gestured: I could see the breakers were already fierce, in a rising wind.

The thought of Summerlee's probability waves intruded into my mind. Could a person in one place become spontaneously transported to another, leaving no footprints in between? Then common sense reasserted itself: even if Summerlee were correct, the effect must be negligible at a non-microscopic scale, or we would notice it more often. I decided to try another tack.

"Is anything known about the man's identity or background?" I asked.

"Yes. He happened to be an acquaintance of one of my men—his name is Andrew Miller. He was of good character, but recently returned from an unsuccessful attempt to emigrate to Australia; he evidently found the life there too harsh for his taste. He had since found slightly unusual employment, as a technician at a local wave research station which two Professors from London have recently set up. Two very brilliant men, I understand, but reputed to be idiosyncratic employers: Challenger and Summerlee, by name."

I resolved not to be distracted by this coincidence. Perhaps novel science would turn out to be involved, but somehow the Australian connection sounded more significant. By chance I had recently dined with a distant cousin also returned from Australia; I tried to remember his anecdotes. Suddenly inspiration struck.

"Inspector, have you heard of the Australian boomerang? Someone could quite well have killed him with that implement! Thrown from, and returning to, its owner's hand, it would leave no tracks, just as we see."

Gregory nodded. "A similar thought occurred to me. But I am told that the returning kind of boomerang is light in weight and thrown purely for sport: the heavy hunting boomerang which brings down prey is made of dense wood and falls to the ground on impact, so I doubt the theory has merit."

I was about to argue, for I was rather taken by my idea, when we were interrupted by a hail. A party of four was approaching. The somewhat comic contrast between Challenger's short boyish figure and Summerlee's sparse frame made them easy to pick out at a distance; they were accompanied by two uniformed policemen. As they drew closer, I was struck by Challenger's haggard demeanor and downcast expression: a greater contrast to his normal cocksure strutting would have been difficult to imagine.

"Could I ask you to formally identify the body, sir?" Gregory asked him quietly.

"Yes: it is Andrew Miller, formerly of my employ. And I am guilty, sir, utterly guilty of his death."

For a moment, I thought I had witnessed a confession to murder, but the Professor continued: "I should have remembered that to a young man, danger is a dimly perceived thing; to impress his peers and elders, any risk may be lightly taken. It was I who put the idea into his mind, and his memories of Australia did the rest."

"Ah, so it involved a boomerang, as I thought," I said.

The Professor looked at me blankly. "Boomerang? What have boomerangs to do with it? It was the waves which killed him: but beyond that it was my pride, my hubris. I thought to stage a demonstration that would surprise Professor Summerlee here. I suggested to Miller how he could play a starring role. But ah, the foolishness of it, to practice alone and out of reach of help. How could I have guessed?"

There were more questions upon all our lips, but at that moment, heavy raindrops began to fall: the storm was upon us. Gregory shouted instructions to his men to bring a tarpaulin so that the body could be covered. The Professors and I left them to it and ran landward. Set in the base of the sea-wall was a small pub, whose sign labeled it The Smuggler's Rest. We piled into its crowded bar, incongruous among rough seagoing types, and found ourselves a table by the window. Challenger gazed bleakly out at the angry sea.

"Come, Professor," I said, seeking to break his mood. "As you said, it is in the nature of young men to be foolhardy and impetuous. You cannot blame yourself for his misfortune. I am sure that the demonstration you planned was of significance. Can you not explain its nature to us?"

Challenger shrugged listlessly. "You would first need to have Professor Summerlee explain the nature of his probability-waves idea," he said.

I turned to Summerlee, who nodded as the waitress brought across our drinks. "The mathematics is complex, but the underlying concept is straightforward," he said dryly. "As I have investigated the tiny world of individual atoms and photons, I have become increasingly conscious of the *randomness* of the phenomena which are observed.

"For example, unstable atoms—like those of which that deadly pair of idols were made—spontaneously split apart, or fission, from time to time. There seems to be no discernible cause for this fissioning, no reason why any one atom should split at a particular instant, even though the average rate is quite constant over time. It is as if some little demon within the atom were throwing dice, and when he happens to roll, say, seven consecutive sevens, then—bang!—that is the cue for the atom to split.

"Moreover, any attempt to pin down precisely the behavior of these tiny things turns out strangely counterproductive. For example, if you want to measure the position of an atom very precisely, you must make it interact with some other entity, after which its momentum—the direction and speed of its motion—is now unknown. The greater the precision with which you measure the position, the more uncertain the momentum becomes, and vice versa. It is reminiscent of trying to bring a magic lantern projector to perfect focus when it is the slide within the lantern—the original photograph—which is blurred. Reality itself is a little out of focus when you stare with eyesight sufficiently keen. Another way to express that is to say that rather than certainties, you can find only probabilities at the very finest level of existence. Probability becomes actuality only when some macroscopic entity, such as a scientist peering through a microscope, makes an observation, and then only fleetingly."

He ignored a derisive snort from Challenger. "This concept of probability waves applies neatly to the two-slit experiment, whether with photons or atoms. The particle is in neither one place nor another until the final measurement is made."

"It sounds a little like a dance performance I saw recently," I said, struggling to understand. "The stage was illuminated by a stroboscope, a light which flashed at regular intervals by means of a rotating shutter. The dancers appeared frozen in different positions on each flash, and the dance was so choreographed that you could never tell by what route a ballerina had moved from one position or pose to another. I am bound to say I did not enjoy the performance much," I said. "You will think me pedantic, but I really prefer the classical opera, where smooth motion is all."

Summerlee nodded. "So do I," he said, "but Nature is as she is, and not designed to be easy for our limited brains to understand."

"So this uncertainty," I said, emboldened, "applies only in the case of individual particles, where there are two distinct paths the thing may choose between, as it were."

Summerlee made to reply, but Challenger interrupted him.

"You have utterly and completely missed the point, Doctor," he thundered. "Really, I would hardly have believed it possible to misunderstand so comprehensively."

He pointed to an adjacent billiards table, on which a game was in progress. "The rules Summerlee describes apply on any scale, and in all situations," he said. "For example, if that man strikes the cue ball at random, as he looks sufficiently inebriated to do, there are several clusters of balls in the middle of the table with distinct gaps between them, through any of which the cue ball could pass while retaining sufficient speed to strike the rear cushion. If the table were in darkness—which would admittedly have to be so pitch black that not one photon of light could escape, nor any other particle: perhaps if the table were enclosed in a double-walled vacuum flask—then the probability of the cue ball hitting a given spot on the far cushion would depend on wavelike behavior: there would be an interference pattern which was a function of the possibility of its taking a multiplicity of different trajectories, like a more complex version of the two-slit experiment.

"Indeed, if the stroke were a vigorous one, all kinds of subsequent histories might potentially unfold upon the table. The cue ball might go left or right; certain balls might end up in the pockets, or not; given balls might or might not momentarily come to rest kissing one another, so that a subsequent impact was transmitted through them. By the time all the balls finally came to rest, their most likely positions would reflect a wave-pattern of all the possible sequences which could have occurred."

At that point, a cat which had been grooming itself upon the establishment's battered piano suddenly took it into its head to jump down onto the billiards table. One of the players had just taken his shot: the cat meowed loudly as the cue ball struck its tail, and whirled in a vain attempt to keep clear of the subsequent reboundings. There was a burst of drunken applause.

Challenger smiled thinly. "Our scenario could even include that cat and its antics," he said. "And since these animals have very thin skulls, in some of the unfolding histories, a ball could hit it at such an angle as to kill it. Summerlee would have you think that all these histories interact to produce the final outcome, observed when we remove the table from its vacuum flask. Just as the interference pattern in the two-slit experiments demonstrates that in a sense the photon or atom passed through *both* slits and interacted with itself in a wavelike way to produce the final outcome, Summerlee presumably believes that the cat was in a sense *both* alive and dead during the time the table was unobserved, to produce the statistics of the outcomes we see on observation. My poor Professor," he said in a tone of mock sympathy, "if only I understood *whose* observation is supposed to have the power to collapse these probabilities into actualities. What if we place Professor Summerlee upon the table, within the vacuum flask, in place of the cat? Is his presence as a competent observer enough to keep wavelike behavior continuously collapsing into actualities through the course of the experiment, so producing statistically different outcomes? Then the superiority of a Professor over a cat is supposed to change the laws of physics the balls obey! What happens, then, if we substitute the cat with an ape-man, just at the dawn of human consciousness? Assuming, of course, that Professor Summerlee ranks above that level, as competent observers go."

The two professors were glaring at each other nose to nose. A silence had fallen in the bar, the patrons able to sense when an argument might be about to break into a fistfight worth a wager. But to my great relief, Summerlee drew back.

"Let us hear your explanation, then," he said contemptuously.

Challenger nodded. "There is a sport popular in Australia," he said, "on parts of the coast where massive waves frequently break upon the beaches. When it is so rough that ordinary swimming is impossible, young men go into the water carrying boards which they call surf-boards. These are made of a wood of low density such as balsa, so that they are light enough to be carried on the shoulder, yet sufficiently buoyant to support the weight of a man.

"The surfer wades out some distance, and waits for a good-sized wave. When one approaches, he scrambles onto his board, and leans his weight forward as the wave hits." He picked up a beer mat, balanced it carefully on the surface of a puddle of beer upon the table—the hygiene of the place was atrocious!—and crooked his fingers upon it to demonstrate the surfer's stance.

"If he balances correctly, he is swept forward upon the leading edge of the wave at great speed. He can vary his course to some extent by leaning one way or the other, but it is mostly determined by the wave. It is an impressive sport to watch, although a dangerous one. The surfer always ends by tumbling off."

"I should think he must often lose his board," I said.

"It is often swept out to sea on the backwash from the wave," said Challenger. "Of course, the next big wave may bring it back in again, so the rider can grab it. But that is a mixed blessing. A wave can also fling the board back while the surfer is still struggling for a footing. If it strikes him on the head—"

"So that is what happened to poor Miller!" I cried.

Challenger nodded. "I fear so. He had just strength to struggle ashore before he collapsed. The receding breakers erased his footprints as the tide went out, and the surf-board was carried away on the waves." He sat with bowed head. "He had boasted to me of his skill on the board. But his physique was not impressive, and I fear he must have exaggerated. Only a novice would be likely to be caught in that way."

"But whatever has this to do with the physics of waves and particles?" I asked. I was glad to have the mystery cleared up, but he seemed to have gone off at a tangent from the original argument.

"Why, my good Doctor, it illustrates that a particle may be *guided* by a wave, while yet being a quite different entity."

There was a dawning light on Summerlee's face.

"Let us suppose that space is permeated by invisible waves, upon which photons and electrons and other particles are guided much as a surfer rides a water wave," Challenger continued portentously. "Consider the two-slit experiment. Suppose the action which emits the electron also starts a wave upon which the electron surfs. The electron is at any given time in one place only, and passes unambiguously through one or the other slit. But the wave passes through both, and as the electron continues to be guided by the wave, its trajectory is affected by the existence of the other slit."

Not for the first time, I was in awe at the Professor's brilliance.

"But how do you explain why a measurement of which slit the electron passes through—an observation—destroys the interference pattern?" asked Summerlee.

"I believe that the measurement knocks the electron off its wave, just as nudging a real surfer even slightly could cause him to lose his balance."

Summerlee sipped his beer, but from his expression he might have been drinking vinegar. "And what of the intrinsic randomness of the atomic world?" he asked. "Where does the uncertainty which prevents exact measurements and predictions being made arise from?"

Challenger smiled triumphantly. "You are familiar with Brownian motion, gentlemen?" We both nodded: I recalled how the jittering of dust particles, due to the random impacts of much tinier air molecules upon them, had convinced my lady patient of the reality of atoms, and so saved her life.

"Dust particles vibrate about in air. The pattern is one of random steps, which by the laws of statistics tend to cancel out over longer periods. The position of the dust particle is well defined at any given instant, but to measure its true velocity—say, the speed of its downward drift under gravity—you must make two measurements well separated in time, or the jittering makes for a large inaccuracy. On the other hand, the position is not well defined over a span of time: the jittering means it is never twice in quite the same place. Is this not remarkably reminiscent of the problem of measuring the position and momentum of an atom?

"I put it to you, Summerlee," Challenger's tone had become almost pleadingly earnest, "that the apparent randomness of the atomic level is merely a consequence of another still more microscopic level beyond the reach of current instruments. Due to incessant waves upon the guiding sea of which I speak, much like the choppiness the open ocean exhibits except in rare moments of flat calm."

The two men gazed at one another, Challenger triumphant, Summerlee expressionless. It came to me that this was a defining moment in the relationship between these two brilliant minds. The next few words spoken could make them forever enemies, Challenger triumphant over Summerlee, or comrades sharing a great discovery.

I leapt to my feet. "Congratulations, gentlemen!" I cried. "I have been baffled by the question, these last months: is the Universe truly particulate, as Professor Summerlee says, or truly wavelike, as Professor Challenger maintains?

"Now I see that you are both correct. The fundamental elements of light and matter are particles: photons and electrons and such. But they all ride upon guiding waves, which determine their destiny. And this subtle truth might never have been uncovered but for the great rivalry between yourselves, which led to both doctrines being defended most vigorously until the truth emerged. Honorable scientific conflict has led to revelation of the facts. Shake hands upon it, gentlemen!"

And, with some initial reluctance the two men shook hands. The rough clientele of the pub, understanding nothing except that a formerly bad-tempered dispute was being amicably resolved, cheered noisily.

THE THREE OF us shared a compartment on the last train of the night back to London. I was still basking in the warm glow of the successful peacemaker when Summerlee said suddenly: "Ah, Challenger, I fancy I have spotted the fallacy in your argument!"

I groaned internally, but Challenger looked up with every appearance of satisfaction.

"Why, congratulations. Do tell, pray," he purred.

"I should have seen it sooner, had I ever witnessed a surfboarder in action," said Summerlee. "Let me be clear: in the case of photons, your guide wave is supposed to travel at the speed of light, is it not? And in the case of, say, an electron moving very close to the speed of light, the wave would have to go at a corresponding speed?" Challenger nodded.

"But that would imply that both photons and electrons sometimes travel faster than light. For by your account they are zigzagging across the wave front. And just as a surfer steering almost along the face of a wave can presumably travel at many times its true speed, so the photon or electron must sometimes travel much faster than light."

"Just as the point where a pair of scissors intersects can move much more rapidly than the closing blades," I exclaimed, remembering the Case of the Faster Businessman. Both men nodded.

"And so according to Relativity the particle must move backward in time, from the point of view of some observers, thus creating an obvious paradox," said Summerlee. But Challenger did not appear disconcerted.

"That is true, in a sense," he said. "I spotted the difficulty, of course. For some time, I myself was troubled by the implication. But, my dear Summerlee, it does not really matter, as long as we never see the particle going impossibly fast. In terms of Doctor Watson's ballet: the impossible motion can be inferred to have taken place between flashes of the strobe light, but it is never directly verifiable."

"That sounds like an ostrich's belief that it can make a predator disappear by putting its head in the sand," said Summerlee contemptuously.

Challenger shook his head. "It is more than that, sir. The point is that the surfing particles can never actually carry *information* faster than light. They are guided by the waves, rather than vice versa, and the mere act of examining a surfing particle knocks it off its superluminal perch. So even if in some inferential sense they sometimes move faster than light, no actual influence or message can be transmitted via them, and therefore no paradoxes can occur."

This sounded somewhat dubious to me, but Summerlee merely nodded, and the remainder of our trip was passed in an astonishingly peaceful silence.

11

The Strange Case of Mrs. Hudson's Cat

So faint was the tap on the door that at first I assumed it was only the wind, and ignored it. But it came again slightly louder, and I opened up to reveal Mrs Hudson's daughter Angela, who appeared to be in some distress.

"Doctor, Mother is most upset!" she cried. "I wonder if you can possibly help?"

My heart sank. Sherlock Holmes really must be London's most trying lodger, and whenever Mrs. Hudson's generous tolerance has finally been exceeded, it generally falls to me to act as peacemaker. This evening Holmes was out, so there was no immediate chance of confronting him with the consequences of his behavior.

"Why of course, if I can," I said. "What seems to be the trouble?"

"It is those confounded stray cats," she said. "Henrietta is in her season, and every tom in London seems to be prowling on our rooftop. They scratch upon the window, and it gives Mother such a start every time. She is in quite a state of nerves."

I was relieved to hear that for once, it appeared Holmes could not possibly be implicated in the problem.

"You wish me to prescribe a nerve tonic?" I asked.

"Oh, no, Doctor, Mother does not approve of those. But the problem is, you see, that the cats come up on the kitchen roof just outside your window here." She indicated the back window. We went across and peered out through the deep yellow fog which had lain over London for the past day. Sure enough, two prowling toms crouched mid-roof, a few yards from the lighted window of Mrs. Hudson's private sitting room. Angela clapped her hands and shouted: the cats vanished as if by magic.

"If it would not be too much trouble, Doctor, to keep an eye out to shoo them off?"

"Not at all," I said chivalrously. "Run along and tell your mother to have no fear. I shall guard our little castle to the best of my ability."

It occurred to me that a good way to scare off the cats would be to wait until they were just outside the lighted window that was their goal, then give them such a fright it would serve as a more effective deterrent. Accordingly, I sought out my revolver and loaded it with a blank charge of powder. With a certain small-boy sense of mischief, I settled myself by the window, which I opened slightly, letting a curl of clammy fog invade the room. I propped the revolver on the window ledge and took up my book, intending to look out from time to time between page turns.

But it proved frustratingly hard to catch any cat red-handed, so to speak. The toms had initially retreated to the lowest eaves of the roof. I peeked out cautiously each time, so that the cats could not see me. I spotted them in various positions, always crouching in mid-slink toward the top. On each successive observation, they seemed sited at random places, as likely to be behind as in front of their previous spots. Yet as soon as I neglected my post for a few seconds, to seek out new reading material, the ensuing commotion revealed both cats had reached their goal. I rushed over to the window, revolver at the ready.

"Good Lord, Watson, what are you doing?" Holmes stood in the doorway.

I turned, slightly shame-faced. "It is all right, Holmes. I was merely about to fire a blank to frighten off some stray cats. They have been troubling our landlady."

"Well, if a stray cat alarms her, a revolver shot should do wonders for her peace of mind!"

I retreated from the window. "I fear I have not been very successful, in any case. Really, Holmes, I would swear there is something spooky or supernatural about cats. Whenever I look out of the window, they are crouched innocently in seemingly random positions. Yet when a moment comes when I do not look, behold, they have attained their goal! It defies the laws of probability."

Holmes snorted. "Hardly, Watson. A cat's senses are refined to detect observers hidden in the shadows. Do you really imagine you can poke your head up silhouetted in a window a few feet away and remain unseen and unheard? Cats merely have the instinct not to give the game away by revealing all that they are aware of. Their movements have been anything but random: they have no doubt been retreating just after you turn away each time, then advancing by cautious bounds."

He sank wearily into a chair. "This reminds me of a problem that has been troubling Mycroft. Pour me a drink, Watson. I am in need of a tonic myself: I have spent the evening listening to him prattle like a nervous woman."

"There is international trouble afoot?" I asked as I busied myself with the decanter.

"No, Watson: certainly not in any immediate sense. Otherwise I could have brought myself to be more sympathetic. The problem is that when you can see as far as brother Mycroft, problems of the distant future which appear remote to you or I are quite vivid to him.

"But this time I really think his imagination has run away with him. He insists that he can foresee problems arising in the new century. He says inexorable forces of history will lead to war, Watson: wars between the Great Powers on a scale never seen before."

I sighed. "Well, it is tragic, Holmes, but war is scarcely a novelty. I cannot imagine any bookmaker giving me good odds that the world will remain at peace for the next half century."

Holmes shook his head. "His worries go deeper than that. He predicts that our growing scientific knowledge will lead to ever more horrifying weapons being developed. His visions of future warfare go beyond even the most nightmarish ideas of that fellow Wells whose scientific romances you so enjoy, Watson. He is particularly afraid of the implications of the new physics whose emergence we ourselves have witnessed: Relativity and Quantum Theory."

"Well the paradoxes of relative motion and the speed of light led to that appalling bomb," I said, shuddering at the memory of a device the size of a football which could have devastated London. "But what is this Quantum Theory?"

"It is just a newly coined name for the nature, both wavelike and particle-like, of light and matter revealed by Challenger and Summerlee's latest studies. Continuous, wavelike entities or probabilities are resolved into discrete, specific quantities—such as electrons and photons whose allowed energy is strictly defined—on observation."

"I can hardly imagine anything threatening arising from a theory which is significant only for describing microscopic entities," I said.

"Well, nothing could have appeared more arcane and abstract than the attempts to measure the speed of light which led to Relativity," said Holmes. "What particularly worries Mycroft is his perception that our understanding of Quantum Theory, and hence of its possible consequences, is as yet very poor."

"But I thought the results of the wave theory had been tested to very high accuracy," I replied, surprised.

"Quantitatively, yes. But then, Newtonian mechanics appeared precisely accurate, until Relativity was discovered. With consequences not just for objects traveling at extreme speeds, but for all those who want to avoid being blown up. Mycroft thinks the true understanding of the theory—its interpretation, or visualization, if you like, of how an entity can be both particle and wave—is most imperfect."

"But I thought Challenger had solved that one neatly, with his wave-riding or surfing picture," I said.

"Mycroft is dissatisfied with it in two respects, Watson. First, there is this question of the surfing particles moving faster than light. It is not altogether clear that this may not lead to paradoxical consequences. Second, there is the question of how the mere making of an observation can be the act that instantaneously reduces an ever-growing sea of possibilities to a single actual outcome.

"One experiment he described reminded me vividly of your problem with the cats. Shorn of its technicalities, the essence is this: You place electrons in a magnetic trap. Initially they reside in the lowest energy state, at the bottom of the trap—like the cats down on the eaves of the roof. If you leave them unobserved, however, an uncertainty in their position develops, as predicted by Summerlee's probability theory. Leave them unobserved for long enough, and some start to escape the trap, like the cats reaching the window.

"But here is the strange thing, Watson. If you periodically observe their positions—merely observe, without disturbing them in any way—then they remain pinned close to their most probable ground state: they never rise up far enough to escape."

I felt incredulous. "You really mean that the mere presence of an aware observer—a human eye at the microscope—has this effect?"

"It is not quite as simple as that. These things cannot be observed with a microscope. The electrons' energy level is detected by firing brief pulses of light. It is the presence or absence of these pulses, rather than an actual witness, which makes the difference."

I snorted. "Really, Holmes, it is exactly like my problem with the cats. Obviously the pulses of light have a physical influence on the electrons; it is nothing to do with some kind of psychic effect of an observer being present."

Holmes smiled. "Your common sense is reassuring, Watson," he said. "That is just what I tried to tell Mycroft. But he insists that these 'observer effects' are robust in their presence in many different situations, quite irrespective of the details of the means of measurement chosen. The presence of any effect which *in principle* could amplify the state of the quantum system studied, so altering the surrounding environment in a way which could subsequently be measured, alters the behavior of that system. It is as if the availability, or the taking, of mere information has an effect different from, and inexplicable by, other known physical laws."

At this point we were interrupted by some commotion downstairs. There were raised voices in the hallway, one of them high pitched. I went to the door. Mrs. Hudson was donning an overcoat and shawl; Angela stood between her and the front door.

"Mother, you cannot go out in this dark and fog! The streets are not a safe place to be now, and you will catch your death of cold besides." She looked up at me appealingly. "Henrietta has gone,

Doctor. She rushed out the front door, which Mr. Holmes cannot have closed properly behind him. Mother is determined to go out and search for her."

Holmes and I were clearly responsible for this situation: Holmes by his carelessness, I for my negligence as a guard.

"Do not trouble yourself, Mrs. Hudson," I called down. "Mr. Holmes and I were just seeking an excuse to take a little walk; why, making it a cat-hunt will give it an object."

Holmes came with rather bad grace, but I was glad of his presence. There is always a slightly eerie feel to London in a pea-souper. Sounds are damped, dying away in a short distance, so that hearing as well as vision is limited to a small hemisphere about you. The occasional stray cat prowled. The ghost stories of Henry James ran through my mind. The fanciful thought came to me that, unobserved, Henrietta might be in no specific place, rather existing as a multitude of ghost cats, awaiting a human sighting to reduce her to reality once more. Was the thought indeed so fanciful? Was this not just what the best minds were saying Quantum Theory implied?

I spotted Henrietta, just as Holmes's muffled call came: "Here she is, Watson!"

"No, she is here. I recognize her positively," I called back. I bent down to pick the cat up, but she snarled at me and streaked away. Nearby I heard Holmes curse. I stumbled over to him: he was nursing a scratched finger.

"Which of us was right?" I asked, peering into the fog.

"Probably neither. What a fool's errand, searching in a London fog on a dark night for a black cat which does not want to be found. A fine task for one who fancies himself a detective!" He snorted. "Let us hope Lestrade never hears of this, Watson, or I shall never hear the last of it. . . . Ah, good evening, Lestrade! What brings you out on such a foul night?"

For a trench-coated figure which had just loomed out of the fog and nearly collided with us was none other than the Scotland Yard man.

"Why, I was coming to see you, Mr. Holmes. But I see you are already off on some errand?"

I was about to explain, but Holmes waved me to silence.

"We were engaged in a little scientific experiment, concerning the effect of fog upon the sight and hearing," he said firmly. "But we were nearly concluded, and the fireside beckons. Come, Lestrade, let us see if we can be of assistance."

On the steps of No. 221B stood Henrietta. Holmes scooped the cat up and presented her to Mrs. Hudson. Lestrade appeared a little baffled at her effusive thanks, but Holmes swept us up the stairs before explanations could be made. Shortly the three of us were sitting comfortably before a well-stoked fire, toddies in hand. Lestrade leaned forward with a touch of embarrassment.

"The problem does not sound too dramatic, Mr. Holmes. It is hardly in the same league as murder or kidnapping. Yet it has the best brains in our fraud department baffled. Our consultant scientist—a man well hardened to criminal inventions!—declares it quite the strangest thing he has come across. Here is the cause of our troubles." Lestrade pulled from his pocket the card reproduced below.

"These cards have recently appeared on newsstands all over London. They are on sale for a shilling each, and constitute a sort of instant raffle. The instructions are printed on the back."

He turned the card over and held it so Holmes and I could read it (see overleaf).

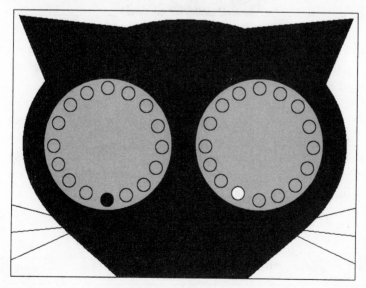

A Winning Card

EXTRA PROFIT RAFFLES

1. The silver eyes of each cat conceal a simple pattern of alternating black and white quarter segments, for example:

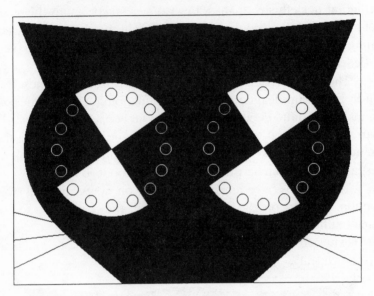

2. The angle at which the pattern is set varies randomly from card to card, but the left and right eye of each cat are identical.
3. On each eye, scratch off the foil at one only of the spots marked near the rim to reveal the color beneath. Each spot uncovered will be either wholly black or wholly white.
4. If the spots chosen differ by only one position, yet the colors revealed are different, a prize of five shillings may be claimed on the spot.

Safety warning:
It is absolutely forbidden to scrape off
more than one spot on each eye!

"I see that someone has won with this card," I commented. Keen to show I was no less alert than Holmes, I counted on my fingers and spoke my diagnosis: "On each circle, there are four places where black and white spots are adjacent, and sixteen possible choices of pairs of adjacent spots. So the chance of winning is four in sixteen, or one in four. For every four shillings spent, on the average you will win five. Why, the company running this scheme must be trying to give away money!"

Lestrade smiled. "Indeed, Doctor, all over this city men as astute as yourself have been coming to the same conclusion, and the cards have been selling like hot cakes. Yet you will scarcely be surprised, Mr. Holmes, to learn that the odds in practice are not quite so favorable. At Scotland Yard, we have tested a large number of cards bought at random, and we find the actual winning chance is only about one in seven. A healthy profit for the vendor is assured."

Sherlock Holmes frowned. "Presumably after each trial, you then scrape away the remaining spots on the card, to check the pattern beneath is as claimed?"

Lestrade coughed in some embarrassment. "Actually, no. The cards have been made by some clever chemist who is determined they shall be quite proof against such investigation. Try it for yourself, and you will see what I mean."

My friend picked up a letter opener and scraped at the uppermost spot on the left eye. At the same instant, the card flashed into flame! In a moment, it was transformed to a small heap of gray ash, from which no detail at all could be discerned.

"We are not sure just how the process works," said Lestrade, "but it seems infallible. In no way can you get more than one bit of information from each eye, and that is whether one spot only is black or white. So we cannot find what pattern is really printed below each, and cannot prove the description given is fraudulent. The real trouble is, scratch our heads as we may, we cannot think of *any* kind of pattern which would give us the results we observe."

I could restrain myself no longer. "Good gracious, Lestrade, I see no mystery here! Obviously, the eyes are colored by some simple rule that yields the results you have seen. Why, I can think of one myself. On six of every seven cards, both eyes are either completely black or completely white. The seventh card has one black eye and one white. Then whichever spots you choose, you will win one time in seven as you have found."

Lestrade smiled. "That was our first hypothesis, Doctor. But there are various tests we can do which are within the rules and do not cause the cards to self-destruct. One is to scrape a spot in the *same* position on each eye, and observe the result. We have done that with hundreds of cards, and in every case we see the same color revealed under both spots. So the claim that each pair of eyes is identical must be true, and there are certainly no cats with one wholly white and one wholly black eye."

Sherlock Holmes spoke thoughtfully. "Although identical, the patterns must evidently be something other than alternating quarters, at least in some cases. Well, the example reminds me distinctly of a beach ball seen end-on. Suppose that in general, the ball is seen at a random angle in *three* dimensions, rather than two. Then the pattern would sometimes appear quite different." And he sketched the example opposite.

"In this case, for instance, there are only two places where black and white spots lie adjacent, and this card would give only one chance in eight of winning. Enough like it could reduce the odds to those you observe. The cunning thing is, the statement on the back could be adjudged strictly true: you would be hard pressed to bring a prosecution. It could indicate the cards' creator is a gentleman of sorts; you might say he is subtle rather than actually malicious."

"Well, Mr. Holmes, you may be accustomed to dealing with gentlemanly criminals, but in my experience the creator of this kind of trick is far from straightforward! In any event, we have already ruled out such a possibility. We tried scraping off spots, on different eyes of course, that were separated by ninety degrees: for instance, the topmost dot on the left eye and the rightmost on the right eye. If the pattern really is as alleged, we should see a different color in *every* such case. Even one counterexample would give us a pretext to bring the rogue in for questioning. But in each case, the color was different."

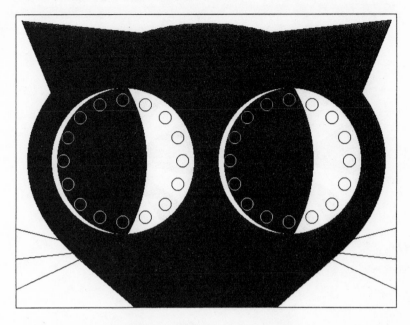

Sherlock Holmes's First Guess

"Then surely that proves the pattern must indeed be alternating quarter circles!" I cried.

Sherlock Holmes shook his head impatiently. "No, Watson, it proves only that it exhibits a certain fourfold symmetry, as follows: take any quarter circle segment, then rotate it ninety degrees and invert its colors—black to white, and white to black—to produce the adjacent segment. Rotate and invert again to produce the third segment, and yet again to produce the fourth and finish the circle. For example, a possible pattern would be as follows." And he sketched the picture overleaf.

"Well," said Lestrade gleefully, "that is a fine hypothesis, but there is just one snag with that particular pattern. It has no less than twelve places where black spots adjoin white, and the chance of winning the lottery with one of those cards would be not one in seven, but three in four! I look forward to hearing of your progress, Mr. Holmes!"

He presented us with a stack of virgin cards for purposes of

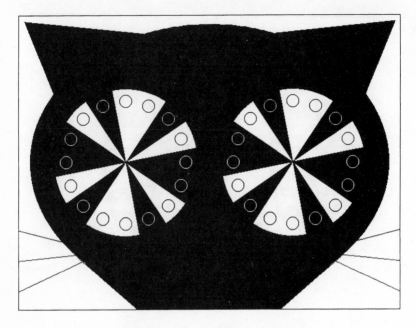

Sherlock Holmes's Second Guess

experiment, and showed himself cheerfully to the door. I was surprised to see my friend remained with furrowed brow.

"Come, Holmes," I said, "there can be no paradoxes in real life. It is merely a question of finding the right pattern."

"Careful, Watson! To be unable to solve a paradox is one thing, but to fail to perceive that one exists is less pardonable. Take it a step at a time. We know that in every case, moving on ninety degrees from a black spot brings us to a white, and vice versa. It is obvious that there must be at least one place in that ninety-degree arc—a traverse of four steps from spot to adjacent spot—where a white spot is neighbor to a black."

"Conceded."

"We know that moving on another quarter-circle would again take us to an opposite color, hence also at least one step across opposite colored spots. Similarly for the third and fourth quarters, which bring us back to our starting point. So we have crossed a minimum of four boundaries between opposite colors. Apparently, we have proved that for any pattern which obeys Lestrade's tests at ninety degrees, there must be *at least* one chance in four of winning, rather than one in seven as found by experiment. A deep mystery indeed!"

When I rose the following morning, it was to behold a red-eyed Holmes still sitting in the rumpled clothes of the day before. Before him was a pile of cards and scribbled paper.

"Thank goodness you are up, Watson. I have a solution, but I need the assistance of a partner to test whether it is correct.

"You know I have always maintained that where other explanations are *impossible*, then the merely *improbable* should be accepted. The only way to explain the observed results, Watson, is that the pattern beneath the spots is not fixed but fluid. Or you might as well say, there is no actual pattern present until you choose to scratch off one of the spots and observe it. The very action of scraping a particular spot on one eye—whether you select the left or right to start with—determines what the pattern of the other becomes. So the question: 'What is the pattern under the eyes?' is initially meaningless: it has no definite answer until an observation is made.

"The mechanism requires some form of communication between the right and left eyes. Since I do not believe in so-called 'action at a distance,' this can presumably be prevented by separating them. Accordingly, I have prepared a large number of cards by cutting them in half. Kindly take this stack of right halves into your room and scrape off one spot from each at random, keeping them in order. I will do the same here with my stack of left halves. When we bring them together for comparison, I do not know just what we shall see. But I will wager my life the result will differ in some way from Lestrade's observations."

It was fortunate no one else was present to take my friend up on his wager, for when we had finished the comparison, the statistics were exactly as before.

"Perhaps," I suggested, "there is some random element at work that does not involve communication between right and left halves. Perhaps the color of a spot is somehow randomly set at the moment it is scratched."

"No, that overlooks the very observation which is at the root of our problems, namely, that in *every* case where the chosen spots correspond, they are of the same color. Any randomness would inevitably result in some counterexamples, unless communication between the halves occurs."

He shook his head. "Watson, I am accustomed to being baffled by complexity. But it is the very simplicity of this problem which confounds me. We have a result which contradicts elementary common sense!"

As we breakfasted, I examined his haggard appearance and took pity on him.

"Holmes, is not this kind of scientific problem really a little out of your line? I am never ashamed to call in a specialist, when in my practice I encounter something I am unfamiliar with."

Holmes brooded darkly for some moments. Then he suddenly snorted with laughter.

"You are right, Watson. It is only my pride which is in the way. I will really hate to be told of some simple, logical solution which I might have spotted. Let us breakfast properly; then we shall call upon one of our scientist friends, and no doubt obtain an explanation."

We walked across the Park to Imperial College, but on arriving at Professor Challenger's rooms, we were told he was engaged in the basement laboratory. Holmes brushed aside an offer to wait, and we descended a set of echoing stone stairs. I pushed open a wide door to reveal blackness, relieved by a faint but eerie blue glow from some large apparatus before us. There came an angry bellow: "Close that door. Confound it, I gave strict instructions we were not to be disturbed!"

A light came on, revealing Challenger and Summerlee standing before the apparatus I have sketched opposite. An enclosing cover had been thrown open to reveal a bench with an electric bulb in the center, the source of the blue glow. At either end of the bench were identical circular glass filters, mounted in such a way they could easily be rotated, and beyond each some lensed device.

Detecting Photons Through Polarizing Filters

"I beg your pardon, Professors. Watson and I have a little problem which has us baffled, which we had high hopes your keen perceptions might solve," said Holmes contritely.

Challenger sighed. I saw that he and Summerlee were hardly in a better state than my companion: both had the look of men who have been laboring all night, and who had not met with a barber's razor for longer than the customary twenty-four hours.

"Well, quite frankly, we are at an impasse here. Summerlee and I have been struggling to duplicate a remarkable experiment just performed in Europe. We neither of us believed the reports of it: but it has been all too easy to reproduce a quite disconcerting result. Perhaps a break for some easier problem will refresh our minds."

Summerlee looked much the less downcast of the two. "You must forgive my colleague's manner," he said primly. "He is a little upset. We have just been demolishing his picture of invisible seas whose choppiness causes local quantum jittering, in the most comprehensive manner."

Challenger waved us to tall stools, and listened with attention as Holmes succinctly but clearly described the paradox of the cards. As he proceeded, both Challenger and Summerlee looked increasingly astonished. Suddenly Challenger slammed a massive fist down on the adjacent table.

"Have you come here to mock us, sirs?" he roared. He took in Holmes's and my expressions, and tossed his head angrily.

"Apparently you have not. It just seems incredible that the problem you describe is so similar to that of the experiment which has been baffling us all night." He indicated the apparatus.

"You know the potential problems with my wave-riding inter-pretation of Quantum Theory—or for that matter with any other I have yet heard? First, that the particles or riders must seemingly go faster than light on occasion. Second, that an observation—the mere obtaining of information about a quantum system—seems to col-lapse it from a superposition of possible states, to a single actuality." We both nodded.

"Well," he continued, "a scientist in Europe has devised an experiment intended to highlight both these problems together. It is perhaps the most ingenious experiment I have ever seen, and the one which has produced the results most difficult to explain.

"A faint source of light"—he indicated the blue bulb—"of a rather special sort emits pairs of photons which travel in opposite directions. The pair have been emitted by a process which ensures they have identical properties, including in particular their polarization."

I coughed. "Forgive my ignorance: I do not recognize that word."

Challenger glared at me impatiently. "It has to do with a prop-erty of photons analogous to spin. For present purposes, imagine each photon is like a little discus, which flies edge on, but may be tilted at any angle. Now imagine the discus approaches a grating, eh?" I nodded.

"If the discus is aligned parallel to the bars of the grating," he continued, "there is every chance it will slip through. If it happens to be at right angles to the bars, on the other hand, it will certainly bounce off. We can coat a piece of glass to behave just like such a grating, with regard to photons: it is called a polarizing filter. These glass disks at each end of the bench are such filters, which can each be rotated independently so that they are at the same or different angles, as desired."

"Now since the two photons are identical, the analogous dis-cuses must always be considered tilted at the same angle as each other. In fact, their angular momentum adds to zero, so if one is spinning clockwise, the other spins counterclockwise, but both lie in the same plane. Now tell me, what happens when each photon reaches its polarizer?"

"Well, if it happens to be parallel to the grating, it passes through; if it is at right angles, it rebounds," I said. "But what if it is at some intermediate angle?"

Challenger nodded. "Then it is a matter of probabilities. The probability of passage is actually given by the square of the cosine of the angle between the grating and the photon's polarization plane. But"—he held up a hand to forestall my protest—"there is no need to follow the trigonometry, to understand the problem.

"Now," he frowned at me fiercely, "it would seem evident that there can be no communication between the two photons. Each must make its own decision, so to speak, whether to rebound or not."

"Why, that is obvious even to me," I said. "In fact, since the photons hit the filters at identical instants, and no kind of signal between them may travel faster than light, it is manifestly impossible that the one collision could affect the other in any way."

Challenger beamed at me. "Why yes," he said. "But then you see, until one or the other is measured, the photons form a single quantum system whose state is indeterminate: it is a mere superposition of probabilities. Or so Summerlee's statistical theory tells us.

"Let us examine what his theory predicts, without getting tangled up in the mathematics. If the two filters are set at identical angles, then the two photons always behave identically. Both pass, or both rebound. You might say it is as if the first photon, on hitting the grating, is twisted either to just the angle at which it can slip through, or just the opposite angle at right angles to the grating, and then the second photon is by some strange force turned to the same angle as the first. So both always behave identically."

"Really, I can think of a much simpler way to explain that," I protested. "Forget your trigonometry: suppose that a photon always passes if its angle to the grating is less than forty-five degrees, and always rebounds otherwise. No strange link is then needed to explain the duplication!"

Challenger nodded. "Very good, Doctor," he said. "And your hypothesis explains a second result which is also observed, namely that when the gratings are turned at right angles to each other, the photons always behave oppositely: one is transmitted and the other rebounds.

"But now," he purred, "what would you expect if I turned the filters at some small angle to one another, say twenty-two and a half degrees, which is just one-quarter of a right angle?"

I pondered. "Well, the result at each would usually be the same, but not invariably," I said. "On my hypothesis, the photons would behave differently one time in four."

"Quite so. Then how would you explain," he roared so suddenly that I shrank back in involuntary fright, "how do you explain, sir, that we get such a difference only about one time in seven?"

The figure of one in seven seemed to trigger some memory from the night before. "I suppose the formula must after all be more complicated," I said feebly.

Challenger shook his head vigorously. "No, sir: no formula, however complicated, can explain the results, unless the photons are somehow in impossible communication with one another.

"Do you not see? *It is exactly the paradox of your lottery cards.* The two photons are like the left and the right eyes of the cat. Setting the two filter angles corresponds to choosing spots to rub off. Setting the filters to the same angle is like rubbing the same spot off each eye: setting them to right angles corresponds to rubbing off spots at right angles, and so forth."

He gave a hollow groan. "It is not only beyond comprehension, it is beyond common sense even if you are prepared to posit the most absurd hypotheses." He glared fiercely at me. "Desperate times call for desperate measures. Summerlee and I have been discussing whether, after all, some signal faster than light might go between the photons.

"Even then, there would be the question of which photon affects which, because relativity tells us the apparent sequence of events is all a matter of frames of reference. For an observer traveling eastward with respect to the laboratory, for example, that left photon reaches its target first." He pointed to the left end of the apparatus. "But for a westward-going observer, the rightmost arrives first. So for one observer, the left photon must first decide its action, and then controls the subsequent behavior of the other. For another observer, it is the right photon which decides, and the left which must follow. It is hardly a logical viewpoint.

"Consider also, this kind of phenomenon affects not just photons, but every particle in every interaction. When I detect some cosmic ray—that is, a charged particle emanating from some distant star—do all those that have interacted with it at any time during the history of the Universe since its creation, everywhere in space, instantly change state? It beggars belief!"

I could feel my head begin to spin. I resolved to return to practical matters. "But what of the lottery cards, then?" I said.

Summerlee waved an impatient hand. "They are clever work, but they merely represent a harnessing of this quantum effect. For example, each side of the card may contain an embedded electron whose spin remains correlated with its twin. Some clever trick of chemistry, no doubt involving a layer of photographic film or something similar, colors the dot scratched off according to a measurement of the electron's spin."

Challenger held up a massive hand. "Listen to yourself," he boomed. "You may well have identified the *principle*, but the *practice* is far beyond the ability of any chemist known to me. It is quite evident that someone is far ahead of us in the harnessing of this new physics, and presumably in its understanding also. Someone who does not wish to cooperate with the scientific establishment, perhaps whose sympathies lie more with criminal activities and even with the Anarchists."

He turned to Sherlock Holmes. "But we have with us the greatest practical investigator in the world. Is it not within your ability, sir, to locate the firm which has produced enough lottery cards to flood all London?"

My friend smiled. Challenger slapped the table. "I look forward to meeting the person who has mastered this strangeness. Warped he may be, but nevertheless deserving of respect. I will never again sleep quite soundly, until we have an answer to this puzzle."

It was as we recrossed the park that a remarkable possibility occurred to me.

"It is a pity that Mr. Rolleman is no longer with us," I said. "For I can see how, believe in Relativity Theory or not, the very lottery cards which litter our living room floor could be used to send messages faster than light—from London to Chicago, or indeed anywhere else. You need only cut a card in half. You take the left half, say, to Chicago, I remain here in London."

"And then?" asked Holmes quietly.

"Why," I said, surprised at his obtuseness, "at some pre-arranged time we each scratch off a spot. You always scratch the top spot, and I will do the top if I wish you to buy and the left if I wish you to sell. If your spot is the same color as mine, I have told you to buy shares."

"And how will I know if my spot is the same color as yours, three thousand miles away?"

"Why, it should —" I paused. "Well then, perhaps we could— no, that would not work either. Come, help me out here, Holmes, I am sure there must be a way!"

Holmes sighed. "No strategy will work," he said. "The problem is that you cannot force any spot on your card to be either black or white. The color of the message you send is outside your control: the correlation between our cards is apparent only when we bring them together for comparison. It is as if the cards' designer has carefully arranged that the internal communication mechanism which makes them work shall be secure from thoughtless tamperers." He seemed to be speaking only half in jest.

"Signaling across time is not permitted, Watson. The Universe is not that strange. But it certainly seems bizarre enough. A somewhat pyrrhic victory for Summerlee, I think: his mathematics have tri-umphed, but produced a result fit to baffle all."

12

The Case of the
Lost Worlds

I SAW ALMOST NOTHING OF Holmes for the next few days. I knew that his search for the Extra Profit Raffles firm was proving far harder than expected, but he was generally out and away before I rose, and back long after I had retired for the night. On the Thursday, however, I returned from my rounds to find him stretched out by the fire, languidly smoking.

He shook his head in answer to my question. "No, Watson: whoever is behind the organization is most infernally cunning. It is almost as if they had some way of anticipating my every action. Today I finally tracked down the company offices, only to find them closed and the bird fled. By chance the rental agent dropped by while I was scouting around. A most striking woman, of Asian appearance but very tall: we attracted embarrassing attention as I spoke to her. She described a group of well-spoken men whom she was mistakenly disposed to trust. Apparently they decamped owing rent, and she is almost as anxious to locate them as I am."

I picked up the previous day's paper, which I had saved for him. "That is too bad, Holmes, for attempts to solve the paradox of quantum measurement which the lottery cards so vividly exploit do not seem to be going at all well. There is a description here of a riotous meeting at the Royal Society held to discuss the matter. It was chaired by our old friend Doctor Illingworth"—Holmes snorted— "and seems to have turned into quite a bear garden.

"After giving Challenger and Summerlee their say, he called for suggestions from the floor. Most were made by philosophers or logicians rather than scientists: all were quite bizarre. For example, one man suggested that the experiment proves that neither free will, nor random chance, truly exist: however arbitrarily you try to make your decision which side of the lottery card to scratch, or which way to turn your polarizer, it has really been preordained for all time. Others suggested that the whole Universe might be subtly interconnected in some way. Several suggested either that the Universe exists only in the minds of conscious observers, or that the presence of a conscious observer defines or melds events in some mystical-sounding way. Some said the whole question of how the effect works is in some sense meaningless, as the mechanism cannot be observed. Some claim the effect is so subtle it is fundamentally beyond human understanding. It all sounds like defeatist drivel to me, Holmes. Mind you, perhaps I do not do the discussion justice: I would not claim to be a philosopher."

Holmes smiled. "And thank Heaven you are not, Watson: I will trust blunt common sense over silver-tongued sophistry any day of the week."

"The bizarrest suggestion of all came from a lady in the public gallery," I continued. "She described the phenomenon of quantum collapse, the resolution of multiple possibilities into a single outcome, quite accurately, and then asked why those present thought such a process had to happen at all! She seemed to imply that rather than one Universe, there might be many, or even an infinite number, in which every possible quantum history gets traced out. The single Universe an observer perceives is, according to her, a sort of consistent illusion, just as in a hubbub of conversation one instinctively picks out a single voice to listen to.

"Apparently some of those on the platform took her quite

seriously. They asked her about the number of such Universes—whether it was truly infinite, whether it increased with time—and why it is the single-Universe illusion arises. But before she could answer properly, Illingworth questioned her scathingly as to her qualifications, and told her the illustrious platform were not present to listen to the ramblings of laymen. She turned on her heel and left the hall. For once, I find myself sympathizing with Illingworth: she does sound like a lunatic.

"But then Challenger told Illingworth that the woman's hypothesis was the first he had heard which explained the experimental facts, without the addition of unnecessary assumptions. Heated words were exchanged. Eventually Challenger seized an ornamental sword which hung on the wall, shouted that it was Ockham's razor, and pursued Illingworth from the platform with it. It was a field day for the reporters present."

"Ah yes, William of Ockham's principle of logical parsimony: do not introduce additional hypotheses beyond the minimum necessary to explain the facts. That medieval English theologian would have made a good detective," Holmes mused.

A point from the newspaper account struck me. "This rental agent of such striking appearance," I asked, only half seriously, "did she wear a blue cloak with a gold clasp in the shape of a crab?"

Holmes sprang up and seized the paper from me. He read swiftly, then flung it down with a curse. "It was her, Watson. Truly, my wits are deserting me. There was a presence, an intelligence, about her which could not be concealed. I sensed it, then ignored my feelings as due to mere attraction. And I let her go! A thousand pities, the more so as I suspect that the one who truly understands these strange matters could well be the brains behind the Anarchists' scientific campaign of terror."

I looked at him sharply. Only once or twice have I ever heard Holmes refer to a woman in a way that implied his emotions were even faintly stirred. Before I could inquire further, however, our door opened to reveal an incongruous sight. A large and portly gentleman stood there, neatly and prosperously dressed, but who was white faced and wheezing as if he had just run a marathon.

Holmes waved him to a chair. "Brandy for our guest, Watson!"

The man was still gasping too much to speak, but he plucked out a business card and held it out for Holmes to take.

" 'Doctor Grainer, Institute of Educational Application,' " he read aloud. "Why, Doctor, I have heard a little of your famous, or should I say notorious, establishment."

The man nodded, either unconcerned at Holmes's slur, or too far gone to protest. "A terrible thing has occurred there," he groaned. "A young man was found dead in his room not four hours ago. He was connected to wires, and had apparently been electrocuted. It is horrible, gentlemen, horrible! This could be the ruin of me."

Holmes's eyebrows shot up. "This sounds most interesting. We must examine the setting before the police have had a chance to disturb too much in their zeal. Summon a cab, Watson: we shall find out the background as we go."

Doctor Grainer's school turned out to be in the Fenlands, and within half an hour we sat aboard a train pulling out of Liverpool Street station. We had a compartment to ourselves, and Doctor Grainer launched into his story without prompting:

"Some twenty years ago, as a young and feckless research student, I made two great discoveries.

"My thesis was on the subject of bitumen, and ways of manufacturing valuable chemicals therefrom. I had the germ of an idea for a novel process. A friend of my supervising professor, a businessman of dubious reputation called Mr. Parkes, took a more than academic interest in the progress of the work. He offered me a substantial sum of money if I could finish the research in good time, some of which he paid in advance.

"Alas, there are many distractions for a young man in a large city, and despite the incentives, my thesis ran ever further over its allotted time. Then early one morning, two large men appeared at my lodgings. They told me that Mr. Parkes wished to see me, and would not take no for an answer.

"I boarded their carriage without too much fuss. They took me not to Mr. Parkes's palatial country residence, however, but to a small inhospitable stone cabin out on the Fens. Reference books, and a copy of my incomplete thesis, were provided. The larger—and uglier—of the two men informed me that I would not be leaving the cabin until my thesis was complete, and that it would go ill for me if I dawdled.

"I am afraid I laughed at him. I explained that creative work cannot be done under duress: indeed, one must be in just the right frame of mind. On some days, a whole morning of coffee and conversation with friends to relax and inspire the brain might be the necessary prelude to a half hour of work of true quality.

"At this point the thug picked me up by the collar and shook me until my teeth rattled. It was then that I made my most astonishing discovery. It is, after all, possible to do creative work, without chatting and drinking coffee indefinitely until the right frame of mind is achieved, if sufficiently strong incentives are provided!"

"A revelation to astound research students everywhere," commented my friend dryly.

"Indeed so. On completing my thesis, and receiving the agreed payment, I decided to invest the money in a sort of sanatorium, designed for those countless thousands of men who are blessed with brains but lack the drive to match. They come to Fairley Farm, and are locked in cells for most of every day, with no distractions save pen and paper. Little luxuries, such as food, must be earned by productive work. A sheet of thesis, a slice of bread: that is our basic barter system.

"The Farm proved immensely popular. Shortly I was forced to raise the fees to a very high level, for the place was becoming oversubscribed. And the institution has run smoothly from that day until today."

He became visibly distressed. Holmes placed a soothing hand on his shoulder. "Tell me about the background of the young man who has died," he said quietly.

"Pemberton? Well, I would have said he was pretty average, for our place. Very bright: he took a First at Cambridge, and then embarked on an ambitious thesis on mathematical philosophy. Three years later, he had done little, and his family had become somewhat weary of supporting him. They are reasonably comfortably off, but not rich. The young man himself had little money. Accordingly, they booked him in to the Farm. The regime worked well on him at first, but recently," he frowned, "his output has not been of the volume I would have hoped for. In fact, now that I come to think of it, the work from most of our guests has been quite disappointing these last few weeks. The modern young man, sirs, do what you will, has become . . ."

He launched into something of a diatribe, which was interrupted only when the train pulled up to a small halt. The journey in the waiting trap was a long one, but as dusk fell we approached a large, low-lying building.

"The isolation helps protect our guests from temptation. Ah, good evening, Kate," said Grainer as a good-looking but solidly built woman came out onto the step. "See, my dear, I have brought the consultant detectives."

The woman snorted. "And you call that a day's work, I suppose, swanning about in London, when a telegram would have done as well? No, there is no need to show them around, I will take care of that. There is work for you to do: the month's account books will not do themselves, just because some drama has occurred."

I noticed, however, that she kissed him tenderly, before leading Holmes and myself along a corridor to a small cell-like room with a high window; its sole furnishings were a desk, a hard chair, and a narrow bunk bed. The body of a handsome young man was slumped over the desk: horrible burns across his forehead showed where the electrocution had occurred. Two wires, affixed to the sides of his head with sealing wax, led out through small holes drilled in the window frame.

"Have you attempted to trace these to their source?" asked Holmes. Mrs. Grainer shook her head. "The local police told us to touch nothing, until the official detectives come from Cambridge."

Holmes nodded. "That is good advice, madam," he said. "I think we will take a little walk outside, in the meantime. I suppose each room has a similar window?"

Outdoors, he led the way along the row of windows, peering closely at each lintel. A pair of fairly conspicuous insulated wires ran from each. Finer wires also ran between adjacent windows like a daisy-chain, well concealed against the wall. All these wires ran ultimately to a little outhouse, which proved to contain a great stack of Leyden jars. The latter had all been wired together in series to produce a high voltage.

If the physical means of the electrocution were crystal clear, the rationale for this extraordinary arrangement remained unfathomable to me. But suddenly Holmes gave an exclamation. He led the way back into the house. Kate Grainer came to meet us: her husband was out of sight, presumably still hard at work.

"Would it be too much trouble to ask for a word with one of your guests?" Holmes asked.

Mrs. Grainer pursed her lips. "I hate to upset their routine. But a brief period of conversation is permitted before lights out. We are only a few minutes early for the normal time."

She led the way to one of the cells, and slid the bolt back. A young man with a weak chin gazed out at us. "Mr. Digsby, these men have come to ask you a few questions about poor Pemberton," she said sharply.

Holmes waited until she left, then turned to Digsby, who looked distinctly uneasy. "The game is up," he said quietly. "I know all about the radio. Tell me the exact instructions you were given for today."

"It was Pemberton's idea, and a harmless prank, really!" the young man cried. Holmes merely waited.

Digsby sighed. "We all are here to work, but sometimes the isolation becomes truly intolerable. On Sundays we are allowed to walk to the village nearby. There is a small radio shop there. Pemberton suggested that we each buy a pair of earphones and a battery. We pooled our cash to buy a single receiver.

"But we arranged the thing so as to avoid excessive temptation. We wired up one battery at a time—the maid was bribed to replace it with another whenever it went flat. To the receiver we connected every pair of earphones in series.

"That way, only when every set of earphones was turned on simultaneously did any current come through. We all agreed to turn our sets on only to receive the news, from one to one-thirty each afternoon. And if any one person—or even all but one—attempted to cheat, it was to no avail. It was really a very disciplined system."

"Until today, no doubt," said Holmes.

Digsby nodded. "Today Pemberton said that as he had invented the scheme, he wanted us to indulge his whim. He sneaked out early in the morning, and slid instructions through each of our windows.

"Mine were that I should listen carefully to the racing results from Newmarket, which are broadcast just after the news. If any other horse than Fiddler's Reach had won the ten-thirty, I was to cross-connect my wires so as to cause a short circuit, being careful not to touch them with my bare hands, as they might be carrying a high voltage.

"Willoughby in the next cell had similar instructions, but concerning the eleven-o-clock and a horse called Long Boy. And so on down the line, I understand. But I give you my word, I had no idea what Pemberton had planned. He appeared in good spirits, even elated: we certainly had no inkling that he might be suicidal."

Holmes nodded. "I believe you," he said. "Thank you. I do not expect we shall have to trouble you again."

Instead of proceeding back to the main house, whence we could hear Mrs. Grainer's nagging tones berating her husband anew, Holmes led the way once more into the cell where Pemberton lay. To my surprise, he started stealthily to go through the boy's pockets.

"What are you seeking?" I whispered.

"Two things, Watson: a betting slip, and a letter. They will not be concealed. Ah, here we are." He pulled from the breast pocket a pink slip of paper, which he passed to me, and a sealed envelope addressed 'To Whom It May Concern.' I stared at the slip in some bafflement: I am not a gambling man, but I could see it did not describe an ordinary bet.

"That is the record of what is called an accumulator bet," Holmes explained on seeing my puzzlement. "It is a bet not on a single race but rather on a series of races, up to a maximum of all those taking place on the course that day, as in this case: five successive races.

"If the nominated horse wins the first race, the stake is multiplied by the appropriate odds, and the whole automatically bet on the horse the gambler has named for the second race, and so on. If any nominated horse loses, the bet is all lost. But if the chosen horse wins in each case, the winnings are multiplied at every stage, without regard to the normal track limits. Thus the gambler stands a very small chance of winning an astronomically large sum. It has been known for a single accumulator win to bankrupt a large bookmaker.

"Do you see, Watson? Pemberton intended that after today he should be either very rich, or painlessly dead. He had arranged that his friends should do the deed, so he would die before he even knew he had lost his bet."

"But why, Holmes?" I cried. "A man who was by no means in desperate circumstances, betting his life against the remotest chance of wealth? It makes no sense."

"Well, as to that, I have a guess," said Holmes. "But I think this envelope was really intended for the Grainers, and should be opened in their presence."

A few minutes later, in a drafty living room, he slit open the envelope, and read aloud:

Dear Sir—

Yesterday I heard a most extraordinary theory. That we may inhabit an infinity of similar universes, diverging one from another at all times as an infinity of possible quantum outcomes all occur together.

The logic has convinced me. I know also from my studies that infinity doubled is but the same infinity as before. Indeed, even infinity multiplied by itself is no larger than infinity. This being so, I have decided to gamble my life at odds of some ten thousand to one.

Of infinite similar copies of myself, only one ten-thousandth will survive: exactly the same number as before. And all will be wealthy beyond my previous dreams.

I am writing to mere figments, to universes which from my point of view will not really exist. But if somebody in some sense reads this, please convey my apologies to my folks, and explain that I took this action not out of any despair, but simple logic. No man has previously thought to take an opportunity such as this.

Yours,

Arthur Pemberton

"One feature of this case puzzles me still," I said as our return train rocked Londonward through the darkness.

Holmes looked up. "And what is that?"

"Mr. Grainer is evidently a clever but somewhat lazy man. It is rather surprising he has been so successful in business. After all, the hired thugs left him alone after he had finished his thesis. Where did his drive come from, after that?"

My friend smiled. "If you ponder what we have seen of Mr. Grainer's domestic arrangements," he said, "I think you will see that he has solved that problem quite satisfactorily."

He put down his newspaper, and looked at me with a more serious expression. "Should you write this little adventure up for your public, Watson, you should perhaps warn them of one or two points.

"First, that the multiple-universes notion is, to say the least, somewhat speculative.

"Second, even if the idea is correct, it is not clear whether there are truly an infinity of parallel universes, or merely a very large number. One ten-thousandth of true infinity is infinity: but one ten-thousandth of any finite number is that much smaller than the original.

"Third, the chance of a suicide device malfunctioning and merely injuring, perhaps crippling, its victim, is probably greater than one in ten thousand—in which case there are more wounded, but surviving, Arthurs than there are wealthy ones.

"Fourth, it is a rather selfish idea, for a multitude of grieving relatives will be created. Unless you take the philosophical viewpoint that universes you can never practicably make contact with do not exist at all, which strikes me as mere solipsism."

I held up a hand to stem the flow. "Holmes, you do my readers an injustice. You wish me to put the kind of warning—'Children:

Do Not Try This At Home'—which appears at the end of juvenile books describing foolhardy adventures. I hardly think any person of common sense needs such a reminder, in this case! But if it makes you feel happier, consider it done."

THE NEXT MORNING, I awoke from a deep sleep to find Holmes shaking my shoulder.

"Step to it, Watson! The game is afoot. The Anarchists have struck, and all good men are needed."

I managed to stagger from my bed and don clothing. It was still black darkness outside. A police carriage was waiting by the curb; nothing else stirred. From Holmes's manner, I had thought a riot was under way at the least, and I told him so rather reproachfully as the horse pulled off. He himself appeared wide awake.

"No, Watson, it is a threat of a more subtle kind. But the police have requested our aid, and we are duty bound to give it."

Halfway down Kensington Road was a police barrier; we could see a matching one about two hundred yards distant. Residents still in night attire were being hustled away. Inspector Lestrade came into sight amid the confusion.

"Morning, all," he said cheerfully. "Well, I hardly see what you can do, Mr. Holmes. It is a simple matter of keeping our nerve while we remove a bomb. Just a little one, this time: a mere few pounds of gunpowder. The only complications are a message clearly intended to confuse us, and a mad Professor who seems determined to interfere with our handling of the matter. Any more fuss from him, and I think he may find himself cooling his heels in the cells for a few hours."

A bull-like bellow in the distance left no uncertainty as to the identity of the mad Professor.

"I am a friend of Professor Challenger's," said Holmes.

"Well, if you can calm him down and get him away from here, be my guest." Lestrade waved us forward. We entered the lobby of an office building to behold a curious sight. A glass tank containing various mechanisms, with a pipe running through the side and a plaque attached to it, stood opposite the door. Challenger stood beside the object, arguing furiously with two policemen. On seeing my friend, he gestured him to approach.

"Thank goodness, Mr. Holmes: you have some experience communicating with these officious public servants, I believe," he boomed. "I am trying to get it through their heads that what we have here is no ordinary bomb, but rather some kind of an intelligence test.

"The plaque explains that the bomb has no timer. But it has a very sensitive trigger. It cannot be safely moved. In fact, even the impact of one single photon of light will set it off! Fortunately, the trigger is in a sealed cavity, which can be accessed only by unblocking the end of that pipe. But the plaque also states that the trigger may in fact have been wedged in position, in which case the bomb is quite safe. However, with no sure knowledge, we must obviously be careful."

Lestrade had come up behind us. "The simple logic is, Professor, that if the bomb is live, it will eventually go off. We cannot keep the area evacuated indefinitely. We can only clear people to a safe distance, and try poking at it. If it is disabled, all will be well; otherwise we must let it explode, then set to tidying up."

"And if you knew for certain the bomb was live, would that be your policy? We are adjacent to several museums containing priceless artifacts," Challenger pointed out.

Lestrade shrugged. "If we were sure the bomb was live, we could take precautions such as sandbagging it. But we can hardly go to such lengths on the off-chance. And obviously with the trigger so sensitive, it is impossible to discover if the bomb is live without actually setting it off."

Challenger snorted. "But that is just what I have been trying to explain to your dunder-headed men here. Given time, I think I could devise an almost perfectly safe test. But even at short notice, I can think of a way to find out if the device is live, with only a chance of detonating it in the process."

"And how do you propose to do that?" Lestrade asked skeptically.

"By firing a photon of light at it."

"But good grief, man, you have just said yourself that that will certainly set it off, if it is live!"

Challenger sighed. "I believe I can explain my plan to Mr. Holmes here. He has at least a basic grasp of the modern sciences."

Lestrade looked at his watch. "I will trust Mr. Holmes's judgment," he said. "But if you cannot convince him in five minutes, then you must clear the area for my lads, or famous Professor or no, you shall find yourself inside the police station." He walked away in disgust.

Challenger drew a pencil and notepaper from his pocket. "Are you familiar with the way that the illusion of a ghost is created on the stage? A sheet of glass is placed at such an angle between the stage and the audience that exactly half the light impinging on it is reflected, while half passes through. The result is that the reflection in the glass and the scenery beyond the glass appear to have equal solidity. An actor in the wings can thus seem to walk across the stage, even passing through furnishings and such.

"The arrangement I intend to use is similar." He drew the sketch shown overleaf. "A photon of light is fired at the angled glass. It has a half chance of being reflected rightward, and a half chance of passing through to strike the bomb trigger, to which is attached a solid angled mirror. To the right of the arrangement, we place a sheet of Doctor Adams's sensitive photographic film."

"Why," I exclaimed, "it is just like an angled version of the two-slit experiment."

Challenger nodded. "Very good, Doctor. If the trigger is wedged, we could fire a large number of photons into the arrangement, and get the familiar banded pattern on the photographic film. If we fire one photon, we may get a dot anywhere on the film, except in the center of the calculated position of one of the dark bands. Interference would of course forbid such an outcome.

"On the other hand, what if the trigger is live, and we fire a single photon?"

"Well, there is a half chance of being blown to glory, I suppose," I said. "But if the photon happens to reflect off the first mirror, it will just make a dot on the film."

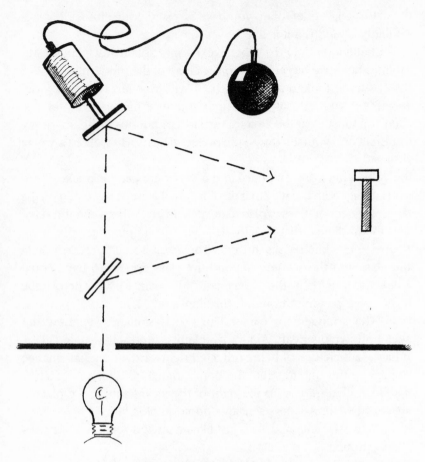

The Impossible Bomb Tester

"And whereabouts on the film?"

"In a random position, anywhere except in the center of a dark band," I said confidently.

Challenger shook his head. "Really, your thought processes are most interesting, Doctor," he said scathingly. "You recall, do you not, that if any attempt is made to measure which slit the photon passes through, the banded pattern disappears. Now would you not call a trigger which activates a bomb a measuring device, which might register on even your senses? If the bomb detonates, we know the photon went through to the farther mirror; if not, that it bounced off the nearer."

I sought to redeem myself. "So if the trigger is live, there can be no potential interference, and the photon is no longer forbidden to arrive in the center of a dark band. It may just as well strike there, as anywhere else."

"Quite so. So let us fire a photon in. It may detonate the bomb, or it may bounce off the first mirror and yet land on the film in a position which tells us nothing; but with a little luck, if it bounces off the first mirror and strikes just where a dark band should be expected from interference, we shall know the bomb is live. Then the bomb will not blow up, *and yet we shall have proved that the trigger was free to move if struck.*"

This seemed quite insane to me. He was claiming that you could tell whether the trigger was firm or loose against the impact of a photon, even when that impact had not actually occurred. There must surely be something wrong with this logic! But my friend was nodding thoughtfully.

We were fortunate to be close to Challenger's laboratory at Imperial College. Within the hour, the equipment he had diagramed was set up. A long wire connected a switch to an electric lamp which could emit single photons. We all withdrew to a safe distance, and lay on the ground with our hands over our ears as Challenger pressed the switch. Nothing discernible happened. There was a snicker from one of the policemen, but Challenger went to retrieve and develop the film. He returned shortly with a white face.

"Inspector," he said very earnestly, "the bomb is live."

It still required some intercession from Holmes to convince Lestrade, but in due course he permitted the area to be thoroughly sandbagged, while the street remained cordoned off despite the chaos caused to the traffic. By evening, every precaution that could reasonably be taken was in place. Challenger had removed the mirror from his apparatus, and took his position with the control switch again. This time, the photon should hit the trigger with no alternative route to take.

I was fully prepared for embarrassment and anticlimax. But the detonation was immediate. The trigger had indeed been live all along. It was a triumphant Challenger who, having verified that the sandbags had successfully minimized the damage, slapped us on the shoulder and bade us good night.

"I CANNOT RID my head of the thought that there is one particular way to look at Challenger's test," I said as we walked home through deepening twilight. "That is, that there were two parallel worlds, which due to the nature of the experiment remained briefly in communication. One was unlucky: the first test pushed back the trigger and set off the bomb. But that bit of information was in effect exported to this version of reality, where we discovered what *would* have happened had we tried the trigger—without actually pressing it."

Sherlock Holmes made no response.

"At least in those terms," I protested, "I am able to understand what occurred. Could I not at least suggest that the many-worlds view is a helpful way to look at things, whether or not it is factually true?"

Sherlock Holmes smiled. "You are cautious, Watson, like Copernicus. It *looks* as if the Earth goes around the Sun, but we will not claim it to be actually the case: we will suggest only that this fiction is a great help in making calculations and predictions."

Suddenly he stiffened; his fingers dug into my shoulder. "Watson, look over there!"

I peered through the gathering gloom. Some distance off, a tall beautiful woman with Asian features was striding along the pavement. A clasp shimmered on the front of her gown. My friend darted toward her, but she noticed him and quickly dodged away into the surrounding throng.

The pursuit that followed would have been comical, had not my friend been in almost pitiful earnest. To every man there comes in due course his *belle dame sans merci*, his siren, his Lorelei—or his Delilah. It was a bond stronger than any curiosity that drew Sherlock onward as we passed from fashionable Knightsbridge through that mixture of wealth and commercial bustle which is theatrical London, on ever eastward toward the seediness of the maritime district. It was quite clear to me, and I am sure to Holmes, that it was no coincidence that the woman had let herself be spotted, and that she was playing him like a fish as she led him onward: but that made no difference to his mesmerism. At last she ducked into a dingy doorway which bore the sign of some kind of club. It was hardly a safe place to follow, but there was no restraining Holmes. We entered, and found ourselves surrounded by opium fumes. The woman had gone deeper into the premises, and we plunged on with swimming heads, to find ourselves in a kind of auditorium already packed with people.

The air was oppressive with the scent of poppies, and I felt my head begin to swim. The master of ceremonies gestured us fiercely to sit down, just as a colorfully attired conjurer appeared on the stage. So thoroughly had she changed in a few seconds that for a moment I failed to recognize the woman we had been following.

She proceeded to perform several conjuring tricks, which, although impressive, were not unique. The audience applauded enthusiastically, but my friend's interest remained fixed on the performer rather than the performance. Then she gestured for quiet.

"So far tonight, my friends, you have seen mere parlor tricks. But now I shall show you something truly amazing.

"Clever scientists have recently guessed that the world we detect with our senses may merely be a cross section through a greater Universe of higher dimensions. From our point of view, our momentary selection of actualities is ever diverging through that multitude: one world becoming many. Give a photon the choice to go up or down, and the two realities in which each opposite took place diverge thereafter.

"My friends, *I can lift that veil between realities.* I can keep touch for a few seconds with that other world as it rushes away from us."

She indicated a large device at the center of the stage, which was draped by black cloth, except for two switches and an electric bulb which protruded from the top.

"When I press this button," she said, indicating the larger switch, "a photon of light is fired at a glass sheet through which it has a one-half chance of passing. If it passes it excites an atom, which in turn triggers this bulb to light. Pressing the smaller button permits the atom to discharge, and extinguishes the light.

"If the photon does not pass, the lamp remains unlit. But the versions of the atom in both possible worlds remain in an entangled state: the one can affect the other. So if at a later moment I permit the atom in the world where the photon passed to discharge, it affects its counterpart in the world where it did not, *and the bulb flickers in that world also.*

"I will prove this to you. I need a volunteer."

Holmes shot to his feet. The woman looked at him and smiled. "No, sir, I do not require a clever volunteer. I need rather a man of utter reliability and trustworthiness, who will not be suspected of complicity in the trick, and whose word is above suspicion. Your companion will serve." To my amazement, she pointed at me.

"Come up on the stage, Doctor. Now follow me carefully. I want you to think of a letter of the alphabet, which you pick at random. Give no sign of what it is. Then I shall press this button.

"If the bulb lights, you must immediately tell me the letter. If it does not, keep it to yourself, and I will tell it to you. Your counterpart in the other world will have revealed it! Let us commence. Think of a letter."

I picked N for north, and nodded. She pressed the large switch on the machine. The bulb lit, and the woman simultaneously clicked a stopwatch which hung on a chain around her neck. "Tell me," she said, and I voiced the N. She studied the stopwatch intently, and at an instant obviously carefully chosen, she pressed the smaller switch, causing the bulb to extinguish. Then she reset both the machine and the watch.

"Think of a second letter," she commanded. I picked O, and nodded. She pressed the large button again, but this time the bulb did not light. She started the stopwatch, and studied the hands intently. A few seconds later, the lamp flickered. The woman looked at me triumphantly.

"Your letter was O," she said.

How could she possibly have guessed that? My mouth hung open in astonishment.

I picked R: the bulb remained unlit, and I kept quiet. "You had R," she told me seconds later.

Proceeding like this, in a few minutes we had spelled out the word *Norbury*. Evidently the precise timing of the signal from world to world—shortly after each split occurred—allowed the letter my doppelganger had revealed to be transmitted. As the last letter was given, my companion sprang to his feet.

"Madam, I wish to examine your device," he said. She shook her head, but Holmes sprang up onto the stage. He was immediately grabbed by two huge men whose role was evidently to keep order in this rough establishment. I went to his aid, but was similarly set upon. I was aware of a sweetly perfumed rag pressed to my face, and the events that followed were hazy.

I CAME TO myself to realize I was sitting upon a hard wooden bench. The cold night air burned in my nostrils like fire. A groan drew my attention to Holmes, who was groggily awakening beside me. With each other's assistance, we staggered to our feet.

"Where are we?" I asked. Holmes gazed about. "Why, in the Euston Road," he said. "Not ten minutes' walk from Baker Street. I suppose our night's adventures could have ended worse."

We walked slowly westward in silence for a short time. At last I could bear the tension no longer.

"Holmes—the events of the last few days—were they really what they seemed to be?" I asked.

Holmes nodded. "Yes, Watson. Everything up to this night, at least. Even Challenger's impossible bomb test was but a version of an experiment which has reliably and repeatedly been performed.

"But tonight's adventure was carefully planned for us. The woman decoyed us to the opium den deliberately, of course. And there, our senses dulled by the fumes, we saw a version of an experiment which has been proposed, but not as yet performed to my knowledge."

"But how could it have been a mere trick?" I said doubtfully.

Holmes laughed. "Ah, the subtle misdirection that mentalists and so-called mediums use to deceive the gullible!" he said. "The word *Norbury*, that you picked at seeming random. The site of one of my most infamous cases. What word are you pledged to whisper in my ear, if you ever see me grow overconfident?"

I felt myself begin to blush. "So parallel worlds are really nonsense," I ventured.

Holmes shook his head. "The idea is merely not yet proven, Watson. You know my maxim that when the impossible has been ruled out, what remains, however improbable, must be the truth. Let us say only that the many-worlds view is the least improbable explanation of the bizarre paradoxes of the quantum world I have seen so far."

We came in sight of University College, with that intimidating inscription on its side gate: 'Let no one enter here, who is not competent in mathematics.'

"I wish this adventure were coming to a tidier end, Holmes," I said eventually. "The matter is still unresolved. But I suppose there comes a point where I, and even you, must yield, and leave the field to those who specialize in philosophy and mathematics and such arcane higher things."

Holmes smiled. "Not so," he said. "It is in the nature of real life to have loose ends, and perhaps in the nature of scientific inquiry that every solution throws up yet deeper problems. The next few years promise to bring revelations, with the likes of Challenger and Summerlee keen upon the trail.

"But never venture to think that such matters can be left entirely to abstract thinkers. Great minds can readily deceive themselves. You have once or twice suggested that even I myself, Watson, have a sneaking preference for the subtle or bizarre, where a simpler explanation might suffice.

"Theoreticians must be kept honest. If a brilliant man cannot explain a matter clearly to his fellow mortals, it is quite likely a sign that he does not really understand it himself. Never fear, Watson; there will always be a role for men of solid common sense, firmly rooted in the practical world. And so there will always be a place in these investigations for the likes of you and me."

Afterword: Paradoxes and Paradigm Shifts

MOST REAL-LIFE SCIENTIFIC INVESTIGATION, like most real-life detective work, involves the diligent following up of clues to confirm and fill in the details of a picture whose broad outlines are already known. But several times in recent history—most notably during a period roughly coinciding with Sherlock Holmes's putative lifetime, from 1850 to 1930 or thereabouts—something far more exciting has occurred, which has more parallels in detective fiction than in normal criminology. The best detective stories feature plot twists which repeatedly cause all the characters and their actions to be seen in a new light. Sherlock Holmes tells Watson how he finds a 'singular feature' which gives the lie to the superficial picture, and reveals a new paradigm governing the actions of those involved. Similarly the story told here is one of scientific *paradigm shifts*. Science is generally supposed to proceed in patient incremental steps. But just a few times in the history of science an experiment has produced a result so paradoxical, so difficult to explain in terms of the accepted order, that the whole framework of current assumptions about the world has to be abandoned in favor of a new, more subtle picture.

The first such paradigm shift was the realization that the Earth itself, always assumed fixed and immovable, was not only rushing through space but rotating at a terrifying speed. Although the concept of a moving Earth had been widely accepted from the time of Copernicus and Galileo, the conclusive proof that it is the Earth, rather than the celestial sphere, which rotates was provided by the pendulum invented by Leon Foucault (1819–1868), which features in **The Case of the Scientific Aristocrat.**

The next breakthrough came with the proper understanding of energy. James Prescott Joule (1818–1889) showed that energy could be converted freely from one form to another, and thus that the notion of phlogiston was nonsense. Joule's initial paper on the subject was rejected by the Royal Society in 1847, but he persisted in spreading his ideas through talks to the general public, and in due course they found acceptance. This advance was of great practical importance. The understanding of the laws governing energy conversion led to the first efficient steam engines and tractors, and so ultimately to the lifting of the burden of physical toil which had been the human race's lot since the invention of agriculture. The detailed events of **The Case of the Missing Energy** are also based on fact. Two divers died tragically in the North Sea in the 1960s, in circumstances very similar to those described. The underwater waves in Norwegian fjords do occur, although they are rare.

Just as the Earth's rotation was suspected long before it could be proved, the notion that matter might be composed of atoms dates back to ancient times. But it took Einstein to realize, early in the present century, that Brownian motion not only demonstrated the existence of atoms but could be used to infer their size, as described in **The Case of the Pre-Atomic Doctor.**

It was Henri Becquerel (1852–1908) who in 1903 shared one of the first Nobel prizes with Pierre and Marie Curie for his discovery of radioactivity by observing the fogging of photographic plates, as described in **The Case of the Sabotaged Scientist.** Investigation of the phenomenon led to the realization that atoms are not truly indivisible, but composed of still more fundamental particles, and that atoms of one element could be transformed into those of another, overthrowing the most basic tenet of nineteenth-century chemistry.

Many investigators took part in experiments to determine the speed of light like those mentioned in **The Case of the Flying Bullets.** The attempt to detect ether drag due to the Earth's motion described by Mycroft was performed by Albert Michelson (1852–1931) and

Edward Morley (1838–1923): for this work Michelson became the first American scientist to receive a Nobel prize, in 1907. It was recognized that the only way to explain the paradoxical constancy of the speed of light in all frames of reference might be to abandon classical notions of the absoluteness of space and time.

Albert Einstein (1879–1955) was the first to clearly see the implications of the new 'elastic space-time' paradigm, which he explored in ingenious thought-experiments. The apparent paradoxes of **Three Cases of Relative Jealousy** are reworkings of Einstein's famous Twins Paradox (which he initially got wrong) and his less well-known Train Paradox, which proved simultaneity to be a meaningless concept. Einstein also demonstrated the impossibility of faster-than-light communication, as sought in **The Case of the Faster Businessman,** and famously proved that mass and energy are equivalent, as revealed in **The Case of the Energetic Anarchist.** The realization that inert matter contains an immense quantity of energy, which may be released by fission or fusion of its atoms, once again demonstrated that subtle shifts in scientific paradigms tend in due course to have immense practical implications.

The two-slit experiment of **The Case of the Disloyal Servant** was first used to demonstrate the wave nature of light by Thomas Young (1773–1829). Einstein photoemission of electrons from metals, and other effects, appeared to contradict this picture early this century by providing evidence of pointlike photons. The paradox led to the construction of a bizarre quantum theory in which the nature of reality (specifically, whether an entity behaves as a localized particle or a spread-out wave) seems to depend on the type of measurement being performed. Einstein commented that trying to understand some of his contemporaries' justifications of this picture was like trying to understand the thought processes of the incurably insane.

The Case of the Deserted Beach (in which Schrödinger's famous cat makes a guest appearance on the billiards table) describes a last-gasp attempt by the late David Bohm and others to present a commonsense picture of quantum reality. The threat to such a description was the problem that, for an appropriately prepared quantum system, a measurement in one place could seemingly affect a far removed particle instantaneously, without regard even to speed-of-light restrictions. This famous paradox, proposed in 1935 by Einstein, Podolsky, and Rosen, was refined by the late John Bell into a series of experiments the most definitive of which was performed by Aspect, Dalibert, and Roger in 1982: Challenger and Summerlee carry out a simplified version in **The Strange Case of Mrs. Hudson's Cat.** The 'observer effect' which Watson encounters when he tries to prevent the cats from invading Mrs. Hudson's parlor is known as the quantum Zeno effect, after the Greek philosopher: it has been demonstrated that beryllium ions can be kept trapped within a magnetic field merely by making periodic observations of them.

The outstanding paradox of quantum theory remains that any given system appears to evolve in a multiplicity of ways, tracing every possible sequence of events—for example, all the possible trajectories of a photon through an optical system—which plentitude reduces to a single specific outcome when a 'measurement,' or interaction with the surrounding environment, occurs. This reduction or *quantum collapse* is an ill-defined event which seems to take place instantaneously when the measurement is made—even if the system comprises elements widely separated in space. The mathematical rules of quantum theory give perfectly accurate results, but there exists no clear picture of the underlying reality. We can all recall being told in school that there was no merit to doing our science homework parrot-fashion—performing calculations as the book tells you, without truly understanding why the result is correct—yet that is exactly the predicament of modern physicists.

Of course the debate about the interpretation of quantum theory continues. To adapt an old joke, if you gather four physicists to discuss the nature of quantum reality, at least five mutually contradictory views are likely to emerge. An extraordinary fact is that the many-worlds view first proposed by Hugh Everett in 1957, in which every possible outcome really occurs in a set of cross-linked or branching Universes, is actually the interpretation which requires the fewest additional assumptions to the proven facts. It has many contemporary supporters, some of whom are reluctant to discuss the matter in public for fear of sensationalization. A physicist at the Santa Fe Institute was allegedly challenged to prove his belief in many-worlds by committing suicide along the lines described in **The Case of the Lost Worlds;** he wisely declined the offer.

The theoretical discussion is perhaps generating more heat than light: I have caricatured it in the meeting which Watson describes at the start of Chapter 12. The current fad is for *decoherence*, a neat mathematical trick which shows how superposed states can decay toward a single outcome. But in the opinion of many, the decoherence concept does nothing to resolve the question of whether superposed states are 'real'—was there in some sense a live cat also present for a while in Schrödinger's box, even if we ultimately open it to find a dead animal?—is there really one Universe or many?

It is reassuring to know that the matter is also being explored by ever more ingenious experiments. (It is a capital mistake to theorize ahead of the facts, Watson!) For example, the 'impossible measurement' technique which Challenger employs to test the anarchist bomb, proposed by Elitzur and Vaidman in 1993, has since been carried out by Anton Zeilinger and colleagues. In fact, the technique has been refined so that the bomb can be reliably tested with a very low chance of detonating it. Other groups are concentrating on demonstrating the reality of superposed states for ever-larger systems, albeit not yet the alive-and-dead cat of Schrödinger's famous allegory.

Does the answer really matter? Some physicists still contend that if the mathematics works for all practical purposes, and there is no guarantee that any experiment can unambiguously reveal which interpretation of quantum theory is correct, we should simply stop worrying. I believe history shows this viewpoint to be wrong. When Copernicus introduced his Sun-centered view of the Solar System, he described it as a mere mathematical convenience, to simplify calculations. The question of whether the Earth 'really' moved or not must have seemed devoid of practical implications. But nowadays we could not understand such a mundane phenomenon as the weather, never mind launch spacecraft successfully, without appreciating the true state of affairs. The transformations of special relativity seemed at first just a useful mathematical device, but they led to the momentous discovery that mass is just a locked-up form of energy. Similarly, the useful mathematical devices of quantum theory may lead to a fundamentally deeper understanding as their implications are worked out.

As the reader may have guessed, my sympathies lie with the many-worlds approach. Several experiments have been proposed to date to test the many-worlds hypothesis explicitly. The final episode of Chapter 12, set in the opium den, is my own extrapolation of an experiment proposed by Rainer Plaga of the Max Planck Institute. In contrast to the other surprising phenomena described in the book, every one of which has been demonstrated in actuality, it is a highly speculative idea, as can be inferred from the context. Even if it proves possible in principle to distinguish between quantum interpretations, and the many-worlds hypothesis is correct (and many physicists would discount each of these assumptions), we will be fortunate indeed if a simple experiment with such clear-cut results is feasible. I have included it, at the risk of being accused of descending into science fiction at the last, to make a very important point. It is that sooner or later, sufficiently clever experiments may cast light on the nature of quantum reality, and the implications could well be important. We should not assume we are done with paradigm shifts yet.

INDEX